SQL

Basics

Datamining Reference Guide
For
Oracle Databases

SQL

BASICS

DATAMINING REFERENCE GUIDE
FOR
ORACLE DATABASES

BRADLEY L SALMOND JR

Frontier Mountain
American Fork, Utah, U.S.A.

SQL BASICS: DATAMINING REFERENCE GUIDE FOR ORACLE DATABASES

Paperback ISBN: 978-1-7366053-5-6
Hardback ISBN: 979-8-8788936-2-6
E-Book ISBN: 978-1-7366053-6-3
Library of Congress Control Number: 2024903068
Edited by: Timothy Schmuhl

Printed in the United States of America 29172-3300

10 9 8 7 6 5 4 3 2 1

Table of Contents

Chapter 1: Introduction

Learning SQL (pronounced "see-kwel") is essential because it empowers you to interact with databases effectively and efficiently, and to extract valuable data. SQL is a powerful tool designed for querying, updating, and managing data within databases.

Datamining, the process of deriving knowledge from data, is a key application of Structured Query Language (SQL). It enables us to uncover patterns, trends, and anomalies in data, leading to improved decision-making, enhanced efficiency, and fraud prevention. As Carly Fiorina, former CEO of Hewlett-Packard, eloquently stated, "The goal is to turn data into information, and information into insight."

What is SQL?

SQL is a programming language used to communicate with and manipulate data in relational databases. These databases consist of groups of tables, that can be connected by like variables in each table. This allows you to easily query and manipulate data across multiple tables.

SQL is a declarative language, meaning the programmer tells the database what they want to achieve, not how to achieve it. This contrasts with imperative languages, where programmers specify the exact steps the computer must take. This declarative nature makes SQL powerful and easy to use for data manipulation. It also makes SQL code more readable and maintainable, simplifies database interaction, and allows for efficient data retrieval and management.

When writing a SQL query, the programmer specifies the desired result, such as a specific set of rows from a table or a value. The database management system (DBMS) then decides how to execute the query most efficiently. The DBMS automatically considers factors like the database schema and indexes to retrieve the data in the most efficient way possible.

Imagine you have a large filing cabinet filled with folders, each containing a collection of related documents. SQL is like the key that allows you to unlock these folders, find specific documents, and organize them in a meaningful way.

Each folder in the filing cabinet represents a table in a relational database. Just as folders have different compartments for different types of documents, tables have different columns for different types of data. For instance, a table of employees might have columns for employee ID, name, address, and department.

The "like variables" that connect tables are called keys. Just like cross-references link folders, keys connect tables, and allow you to link related information together. For example, the EMPLOYEE_ID column in the ORDERS table can be linked to the EMPLOYEE_ID column in the EMPLOYEE table, showing which employee placed each order.

Why Learn SQL?

SQL isn't just a powerful tool for data analysis, it's a versatile, portable skill that empowers you to extract meaningful insights from large and complex datasets. It is relatively easy to learn and can be used to do a variety of tasks in databases. Some of the things you can do with SQL include:

- Create, modify, and delete tables
- Insert, update, and delete data
- Retrieve data from tables
- Join tables together
- Create views
- Write stored procedures

Who Should Learn SQL?

SQL is indispensable for anyone who works with databases, including database administrators, developers, and analysts. It's a highly sought-after skill across various industries, including technology, finance, healthcare, retail, and manufacturing. Companies rely on people who are highly skilled in SQL to analyze and manage large amounts of data for decision-making, making SQL-proficient individuals' valuable assets.

SQL proficiency opens doors to a wide range of job opportunities, including data analyst, business intelligence analyst, data scientist, database administrator, and software engineer. These roles are in high demand and offer attractive salary packages. SQL proficiency can also accelerate your career advancement, allowing you to move into more senior positions with higher salaries and greater responsibilities.

As data becomes increasingly central to business operations, the demand for SQL skills is expected to continue growing. Because SQL is a standard language that is supported by all major relational databases, your SQL skills are transferable across industries and technologies, making you a versatile and adaptable candidate.

The following are some examples of how SQL is used across a variety of occupations:

- A data analyst can use SQL to query a database of customer transactions to identify trends and patterns.

- A marketing manager can use SQL to segment a customer database and target specific groups of customers with targeted marketing campaigns.

- A software engineer can use SQL to build database-driven applications.

- A data scientist can use SQL to prepare and clean data for machine learning models.

Chapter 2: Database Structure

What is a relational data base?

A relational database is a collection of tables that store data. Tables can be related to each other through foreign keys, which are columns that reference the primary key of another table.

These keys are data points that relate to one or more tables in the database. Each table has a predefined set of columns which represent different attributes of the data, and a set of rows, which represent individual records of data. This provides a way to store related data in a way that is both efficient and easy to query (see Figure 3: Relational Database on page 12).

Imagine navigating a vast filing cabinet overflowing with folders and documents. Finding a specific paper amidst this chaos can be a nightmare. But what if each document had a unique, never-changing ID, like a personal fingerprint? This is the magic of a primary key in the realm of databases.

Think of each folder as a table, and each document as a row. The primary key, akin to that unique document ID, is a special label that sets every record apart. No two documents can share the same ID, and every document must have one to prevent document doppelgangers and ensure swift retrieval. It is their one-of-a-kind passport in this organizational world.

This key element ensures every document stands alone, never to be confused with its neighbors. No two documents wear the same fingerprint, and none can exist without it. This uniqueness ensures that you can pinpoint any document instantly, even if it's buried deep within subfolders.

But the benefits go beyond simple retrieval. Primary keys prevent accidental duplicates, keeping your data clean and accurate. They also act as bridges between folders, allowing you to connect related information and build a web of knowledge within your database.

Parts of a Database

1. **Data element:**
   The smallest unit of data in the database. It is a single value, such as a name, a number, or a date.
2. **Row:**
   A row in a database is a record of data that is stored in a table. A row is made up of columns, which are the individual pieces of data that are stored in the row.
3. **Column:**
   A group of related data elements. For example, a column might contain the names of all the students in a class.
4. **Table:**
   A collection of related columns. For example, a table might contain all the information about students in a class, including their names, email addresses, phone numbers, or birthdays.
5. **Schema:**
   A schema encompasses a cohesive set of interconnected tables, creating a logical structure that aligns with specific aspects of your business or application.

For example, one schema might house all student-centric details, cataloging student information, class enrollments, and academic performance within the class, and grade tables. Meanwhile, a separate schema might contain tables with the details of faculty, documenting their courses, schedules, professional milestones within the class taught, and tenure.

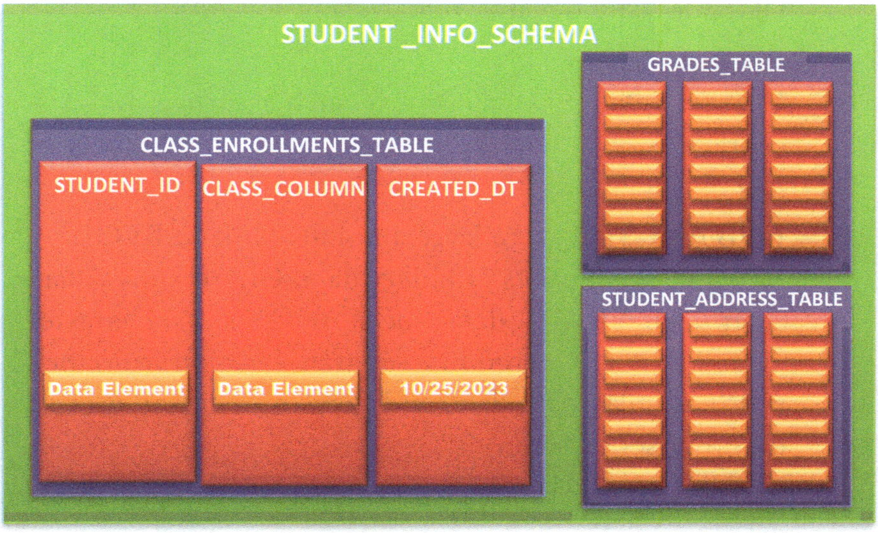

Figure 1: Schema

Data Types

Data elements are the fundamental building blocks of data, serving as the smallest units of information that convey meaning. They are like the individual pieces of a puzzle that, when combined, form a complete picture of data. Data elements can be qualitative, representing descriptive attributes like customer names or product categories, or quantitative, representing numerical values like sales figures or inventory levels. Each data element has a unique name, a clear definition, and a set of allowable values, ensuring consistency and accuracy in data collection and processing.

1. **Numeric:**
   Numeric data types store numbers, such as integers (numbers without decimals), floats (numbers with decimals approximately using scientific notation), and decimals (the exact figure).

2. **Datetime:**
   Datetime data types store dates and times.

3. **Character:**
   Character data types store text, such as strings and characters.

4. **LOB:**
   LOB data types store large amounts of data. BLOB stores large amounts of binary data, CLOB stores character data, NCLOB data type stores Unicode character data. These are stored in a separate location from the rest of the database data, and they are accessed using a different set of methods.

5. **Unicode:**
   A Unicode character is a unique code point that represents a character from all of the world's writing systems in a consistent way. Unicode characters are encoded in a binary format that can be stored and transmitted in computers and other devices.

6. **Binary:**
   Any data that is stored in a format that is not human-readable, such as images, audio, video, or compressed files.

7. **Spatial:**
   Spatial data types store geographic data, such as latitude and longitude coordinates.

8. **Object:**
   Object data types store complex data structures, such as nested arrays or objects.

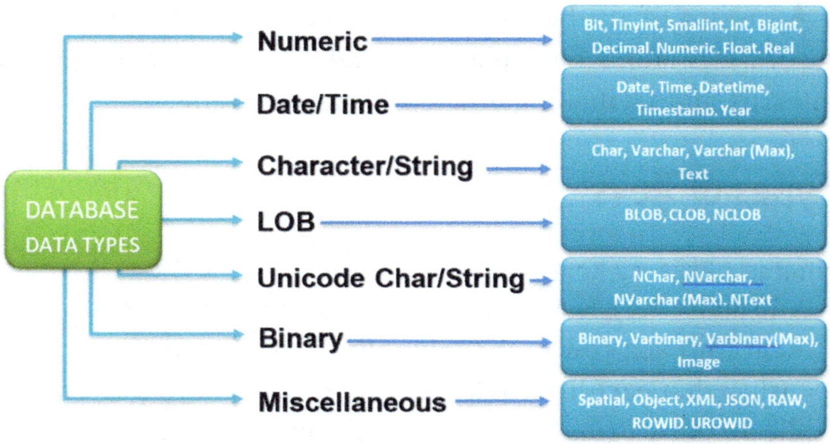

Figure 2: Data Elements

NULL Values

When there is no data in a column, the absence of data is called "null." NULL is different from a zero value (0), or an empty string as it signifies the non-existence of any data or value in a particular field.

You might see nulls in a data set because information might not be available when a record is created, some fields aren't required for every record, or a record might not have a corresponding entry in a related table.

Chapter 3: Keys

In a relational database, keys are specific attributes or groups of attributes that help to uniquely identify each record in a table. They play a crucial role in maintaining data integrity and facilitating efficient data retrieval.

Types of Keys

What is a Primary Key?

A Primary Key is a column that provides a unique data element for each row in the table. Primary keys are used to ensure that each row in a table is unique and to prevent duplicate data. Primary keys are typically assigned automatically by the database when a new row is added to the table. However, you can also specify a primary key manually. Columns with primary keys will never have null values.

What is a Foreign Key?

A foreign key is a column in a table that references the primary key of another table. It establishes relationships between tables in a relational database.

- Foreign keys are essential for enforcing referential integrity. Referential integrity is a set of rules that ensures consistency between tables. For example, if you have a table of customers and a table of orders, you might use a foreign key to ensure that the CUSTOMER_ID column in the ORDERS table references the CUSTOMER_ID column in the

CUSTOMERS table. This would ensure that each order has a valid customer associated with it.

- Each table in a database should have a unique identifier for each row. This unique identifier can be either a primary key, which is a column or set of columns that uniquely identifies each row in a table, or a foreign key.

- If the two tables you want to link together don't have keys that you can link together, you have to go through other tables to link them. You can link tables together like a chain as long as each table has a one-to-one relationship with the tables to which you are linking.

For example, using the diagram on the next page, say I wanted to write a query that told me which professors had the most students receiving Financial Aid of a specific type, which required me to get information that was only in the FINANCIAL_AID table.

You would start with the PROFESSORS table and join it to data in the COURSES table using the COURSE_ID key (because that key is in both tables). Then, you would join the COURSES table to the STUDENTS table using the STUDENT_ID. Then you would join the STUDENTS table to the FINANCIAL_AID table using the FIN_AID_ID key.

One could perhaps go through the GRADES table instead of the COURSES table as the STUDENT_ID and COURSE_ID exists in both tables. However, if they did take that route, they run the risk of missing data because a student may not yet have received their grade.

Chapter 11 goes into the detail of the syntax one would use in a query when joining tables.

The diagram below represents a simplified example of what a relational database might look like with a variety of schemas, tables, and primary keys in each table that tie into the same key in another table.

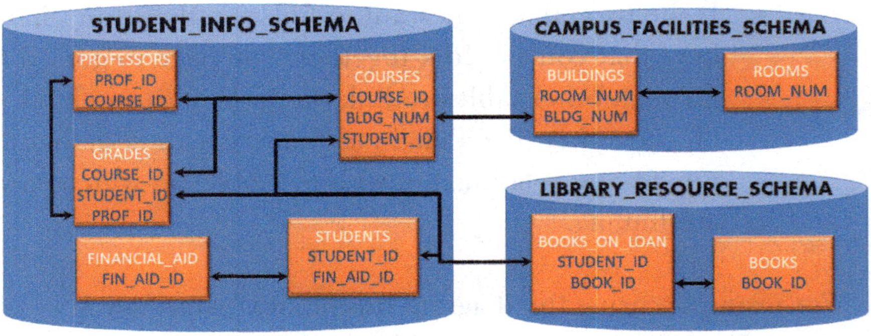

Figure 3: Relational Database

What is a Composite key?

In some cases, a single column may not be enough to distinguish individual rows, a composite key ensures that each record possesses a distinct identity. A composite key is formed by combining multiple columns that, individually, wouldn't form a unique key. A composite key functions as a primary key, even though the data types of these columns may vary. In some cases, a composite key can be formed by aggregating all the foreign keys in the table.

For instance:

- Suppose you had a table that records a weekly log of work output for every employee in the company. Every entry has the Sunday date for the week that represents the work in which the work output occurred. If every employee worked the full year every employee would have 52 entry's in this table.

  In this example the Sunday date is not a unique number neither is the employee_id, however an employee can only have one entry in any given week so combining the employee id and the Sunday date into a composite key would create a unique identifier for each row.

- Creating your own unique keys by making a composite key can be a bit more complex than using a single column as a primary key, but it can be necessary in some situations.

Chapter 4: Table Relationships

The concepts covered in this chapter are fundamentally crucial for all query writers to understand. Table relationships form the backbone of relational databases, defining how data entities interact with each other.

Different types of relationships (one-to-one, one-to-many, many-to-many) cater to diverse data structures and ensure data integrity. Understanding and mastering these relationships is key to building efficient and functional databases and will empower you to query, design, and manage complex databases effectively.

Table Relationship Types

One-to-One

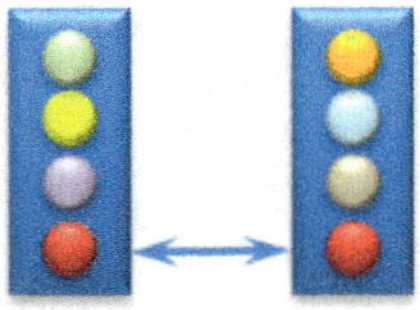

Figure 4: One-to-One relationship

A one-to-one relationship between two tables in a database is a relationship where each record in one table is associated with (or uniquely identifies with) only one record in the other table.

For example, a table of employees might have a one-to-one relationship with a table of employee social security numbers, or a table of students email addresses might have a one-to-one relationship with a table of student transcripts in a given year.

To create a one-to-one relationship between two tables, you need to have a foreign key in one table that references the primary key of the other table. The foreign key must be unique, meaning that no two records in the table can have the same value in the foreign key column.

One-to-Many

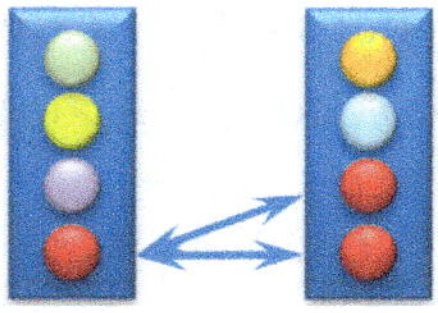

Figure 5: One-to-Many relationship

A one-to-many relationship between tables in a database is a relationship where one record in one table can be associated with multiple records in another table. This is the most common type of relationship in a database.

For example, a table of students might have a one-to-many relationship with a table of courses. If the student were taking five classes he would have five rows in the courses table. In the student table he would have one row. Linking these two tables together would create a one-to-many relationship.

Or the AUTHOR_ID column in the BOOKS table would be a foreign key that references the AUTHOR_ID column in the AUTHORS table. This would ensure that each book has a valid author associated with it, but an author can write multiple books.

Many-to-Many

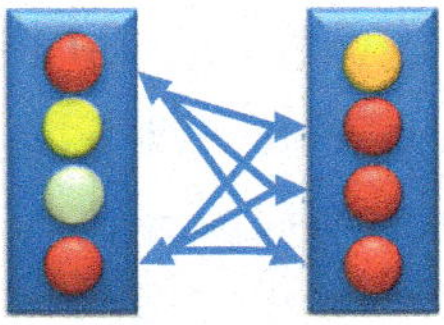

Figure 6: Many-to-Many relationship

A many-to-many relationship is a type of database relationship where one record that is duplicated in one table can be associated with multiple records in another table, and vice versa.

Imagine you have a table called "COURSES" that lists all the courses offered at a school. This table would have columns like:

Course ID: A unique identifier for each course.

Course Name: The name of the course.

Instructor: The name of the instructor.

Student ID: The ID of the student enrolled in the course.

A student can be enrolled in multiple courses, so their ID would appear in multiple rows of the "Courses" table. For example, if a student takes 3 classes, their ID would appear in 3 rows on this table.

The second table called "ATTENDANCE" records the attendance for each day of class. This table would have columns like:

Course_ID: a unique identifier for each course

Attendance_ind: an indicator denoting the attendance for the day

Student_ID: The ID of the student enrolled in the course

If the semester lasted 180 days there would be 180 rows for each student in the ATTENDANCE table. If you were looking at the student that had taken 3 classes you should expect to see 540 rows in the ATTENDANCE table (3 x 180 =540) for that student.

You receive a request to pull a student's attendance records for every course they attended. You decide to join the COURSES table to the ATTENDANCE table using STUDENT_ID, a key that exists in both tables multiple times in each table.

Tying the two tables together by STUDENT_ID would tie every class the student took in the courses table to every attendance record for that student in the attendance table. When the 3 class records attach to 540 attendance records each the resulting data set will return 1620 rows. This is a many-to-many relationship.

Here's why it gets messy:

Each of the 3 courses gets paired with all 540 attendance records, even though they might not belong together.

It's like a matching game where you have 3 names (courses)

and 540 photos (attendance records), but you're matching every name with every photo, even if they don't match. This creates the jumbled list of 1620 combinations (3 courses x 540 attendance records), making it hard to see which attendance records belong to which course.

In most cases you will want to avoid creating many-to-many relationships in your queries as this can cause a Cartesian product. A Cartesian product is the set of all possible ordered pairs of elements from two sets.

Why are Table relationships important?

Not understanding table relationships can be the biggest stumbling block for most new query writers, and sometimes even veteran query writers if they are not attentive. Understanding what kind of relationship exists when you are joining tables together is crucial to effective query writing.

There have been many who have joined two tables together thinking they were getting a one-to-one relationship just to find out later that they had gotten a one-to-many by mistake and had published the wrong data. There have also been occasions where the individual had known they had a one-to-many relationship and tried to turn it into a one-to-one through the where clause, but hadn't factored in all the different data types and still had a one-to-many.

For instance, if you were to join a product table with a supplier table, then expect each product to have only one supplier. However, upon closer inspection one would find some products are sourced from multiple vendors and this wasn't discovered in your initial review of the data.

Inventory reports or purchase orders generated based on this query would be inaccurate, potentially leading to overstocking or understocking of certain products. This could create delays in fulfilling customer orders or incur unnecessary storage costs.

It is important to test your queries carefully to make sure that they are returning the correct results. This will help you to identify any errors in your understanding of the relationships between the tables.

Table relationships are crucial because they enable you to connect data from different tables. This can be useful for a variety of purposes, such as:

- **Querying data:**

  Table relationships allow you to query data from multiple tables simultaneously, helping you discover connections between different data elements.

- **Creating reports:**

  With table relationships, you can create reports that display data from multiple tables. This can be helpful for summarizing data and making it easier to understand.

- **Enforcing data integrity:**

  Table relationships enforce data integrity by ensuring consistency between data in two tables. For example, in a one-to-one relationship, each record in the first table must be associated with exactly one record in the second table.

Chapter 5: Basic Parts of a Query

A query is a statement posed to a database to retrieve or manipulate data. It serves as a request for specific information from a database, enabling users to gain insights and make informed decisions, and provides a structured syntax for interacting with databases.

The goal of writing a query is to extract meaningful information from a database.

Each part of the query, if used, needs to be listed in the query in the order shown below.

```
SELECT column_names
FROM table_names
WHERE first filter or condition
AND additional filters or conditions
GROUP BY column_names
HAVING filters for aggregate functions
ORDER BY column_names or column numbers
FETCH first 10 ROWS ONLY;
```

Of the above items, only **SELECT** and **FROM** must be used in every query. The other items are optional depending on what you are trying to accomplish with your query and what kind of results you are trying to return.

SELECT (column_names)

This is primarily column names, separated with commas. It specifies which columns you want to include in the query results.

FROM (table_names)

This is primarily table names.

- These tables are the source of your data.
- All tables need to be joined together using table keys.
    - Tables may be joined in the from clause using the ANSI join types *(See pages 51-59)*.
    - Through joins tables can be linked together like a chain.
- Queries can be used in the **FROM** statement to create your own table (more on this in chapter 13).

WHERE (first filter) / AND (additional filters)

This is where you specify how you want to filter the data.

- The **AND** keyword is used for additional filters after the first **WHERE** filter.
- Tables may be joined in the **WHERE** clause using the ORACLE join types *(See pages 60-61)*.

GROUP BY

The **GROUP BY** clause is used to group rows together based on common values in one or more columns. It is used with aggregate functions in the **SELECT** statement (COUNT, MAX, MIN, SUM, AVG) to group the result-set by one or more columns.

- **GROUP BY** will always contain the same columns in the **SELECT** statement that are not aggregate functions and without the aliases.

- **ROLLUP:** In addition to the regular aggregation results we expect from the **GROUP BY** clause, the **ROLLUP** extension produces grouped subtotals.

The following are two examples of a **GROUP BY** with a **ROLLUP** statement included:

```
SELECT q.yrqtr, SUM(amount)

FROM SCHEMA_1.TABLE q

GROUP BY ROLLUP(q.yrqtr);
```

or

```
SELECT q.yrqtr, SUM(amount)

FROM SCHEMA_1.TABLE q

GROUP BY q.yrqtr WITH ROLLUP;
```

The resulting data set from these queries might look like this:

QTRYR	SUM(AMOUNT)
1_2023	13,001
2_2023	10,002
3_2023	12,001
4_2023	8,004
	43,008 ← **ROLLUP** produces this figure.

The ROLLUP command creates the last line in the result set which is a sum or rollup of all the numbers in the column.

HAVING

This is similar to the **WHERE** clause but is for aggregate functions. It is a set of conditions that you want your aggregated results to meet and is used to filter the results of the **GROUP BY** clause.

Example: A music streaming service wants to analyze its users' listening habits to identify popular artists and genres. They have a database table named LISTENING_HISTORY that stores the user ID, artist name, genre, and the number of times each song was listened to.

They want to find the top 10 most popular artists based on the total number of times their songs were listened to. However, they only want to consider artists with at least 10 songs in the database.

We can use a **HAVING** statement to filter the results based on the total number of songs per artist. Here's the SQL query:

```
SELECT h.artist_name, SUM(listen_count) tot_listens

FROM SCHEMA_1.listening_history h

GROUP BY h.artist_name

HAVING SUM(listen_count) >= 10

ORDER BY 2 DESC;
```

This example demonstrates how the **HAVING** statement can be used to refine the results of a **GROUP BY** operation based on specific conditions. It allows us to focus our analysis on the most relevant data and gain valuable insights from the music streaming service's user data.

The resulting data set from this query might look like this:

ARTIST_NAME	TOT_LISTENS
Billy Noel	35
Frank Natra	22
Elvis Partridge	18
Bob Dilly	11

ORDER BY

This is used to order the results by one or more of the columns in the **SELECT** statement.

- You can list the column number (e.g., "**ORDER BY** 1" to order by the first column).

- You can list the column name you want to order by (e.g., "**ORDER BY** c.P_KEY2" to order by the P_KEY2 column).

- To order by multiple columns list each column separated by a comma that you want to order by.

For Example:

```
SELECT c.key, c.name, c.city

FROM TABLE_1 c

ORDER BY 3, 2, 1
```

The results of this query would appear as follows:

KEY	NAME	CITY
457	Patrick	Denver
458	David	Dover
456	Able	Junction
455	Sam	Junction
459	Sam	Junction

- **DESC**: **ORDER BY** will sort ascending unless **DESC** is added to the end of the **ORDER BY** statement (eg. **ORDER BY 1 DESC**)

For Example:

```
SELECT c.key, c.name, c.city

FROM TABLE_1 c

ORDER BY 3 DESC, 2, 1 DESC
```

The output of this query would appear as follows:

KEY	NAME	CITY
456	Able	Junction
459	Sam	Junction
455	Sam	Junction
458	David	Dover
457	Patrick	Denver

FETCH or LIMIT

These will let you grab only the number of specified rows, which can speed up a query that has to grab a large amount of data.

The syntax for fetch is:

FETCH FIRST 10 **ROWS ONLY**.

The syntax for limit is: **LIMIT** 10.

Where 10 is the number of rows you want to capture. You will get the first 10 rows.

Semicolon ";"

In SQL, the semicolon (;) is used to denote the end of a query. This is useful when you have multiple queries on the same tab, as the semicolon will tell the database tool to execute each query individually.

Depending on the software you are using, if you run the script with a semicolon at the end of each query, the script will start running starting with the query where your cursor is located and will proceed to execute each query from that point forward.

A Basic Query

The most basic and commonly written query is the following:

```
SELECT *

FROM SCHEMA_3.TABLE_A1;
```

The * indicates you want to select everything from table_a1.

When two tables are involved and you only want to see the data from one of them you can use the following:

```
SELECT  a.*

FROM SCHEMA_3.TABLE_A1 a, SCHEMA_3.TABLE_A2 b

WHERE a.P_KEY6 = b.P_KEY6;
```

The (a.*) indicates you are selecting everything from TABLE_A1 only (a. is an alias, aliases are covered in the next chapter).

Chapter 6: Aliases

Aliases are assigned in the **SELECT** statement or in the **FROM** clause, but not in the **WHERE** or **AND** clauses. They serve the purpose of giving a column or table a different name. This can be valuable for enhancing the readability of queries or simplifying references to columns or tables in subsequent clauses.

Using Aliases in the SELECT Statement:

When used in the **SELECT** statement the alias changes the name of the column.

- Column aliases are typically separated from the column name with a space. However, some prefer to use the **AS** keyword to make it more explicit that what follows **AS** is the alias. While **AS** is not mandatory, it is a recommended practice to use it to enhance query readability.

- To use the space, you can list the column name, followed by a space, and then the alias.

For example:

```
SELECT STATUS_CD employer_type
```

or

```
SELECT STATUS_CD AS employer_type
```

In this example if the alias were not used the column would be labeled STATUS_CD. With the alias in place as shown the column will be labeled EMPLOYER_TYPE.

- Aliases typically cannot have a space in the name, thus the underscore "_" character is generally used. However, if you would like to use a space or any character that is not usually allowed in an alias you may use them by wrapping the alias in double quotes (").

  For example:

```
    SELECT STATUS_CD "employer type"
```

or

```
    SELECT STATUS_CD AS "employer/*type"
```

  This example would display the column name as EMPLOYER TYPE and EMPLOYER/*TYPE respectively.

- When long column names are in the table it may be more convenient to use aliases:

  For example, if your table contains these columns: vehicle_make, vehicle_model, percent_of_total_car_sales

  Then you may want to give them aliases like in the following query:

```
SELECT vehicle_make AS make

     , vehicle_model AS model

     , percent_of_total_car_sales AS pcnt_Sales

FROM sales_ledger
```

This would yield the following results:

MAKE	MODEL	PCNT_SALES
Ford	Fiesta	5%
Chevy	Avalanche	10%
Jeep	Wrangler	15%

Using Aliases in the FROM Clause:

When used in the **FROM** statement, the alias provides an abbreviation for the table name. This can be useful for making queries more readable or for making it easier to refer to tables in subsequent clauses.

- List the table name, followed by a space or **AS**, and then the alias.

```
        FROM SCHEMA_2.TABLE_A1 s
 or
        FROM SCHEMA_2.TABLE_A1 AS s
```

Once the alias is assigned you can use it instead of typing out the full schema and table name when listings columns in the SELECT or WHERE clauses. Simply use the alias "s." (e.g., s.P_KEY1).

```
SELECT s.order_id, s.sale_price

FROM schema1.sales_log s

WHERE s.created_dt > '2 JUL 2020'
```

- Use the aliases in the FROM statement before the column names in the SELECT and WHERE clauses to specify which table the column originates from (eg. s.created_dt, s.customer_id)

Chapter 7: Comments in SQL

Comments are non-executable statements inserted into SQL code to add notes, explanations and improve readability. They are essential for documenting your code and communicating your intentions to others who might read it. Comments can also be used to explain complex logic or provide information about data sources. They provide explanations, annotations, and instructions within the script, enhancing readability and understanding for both the writer and others who may need to read or modify the code.

Benefits of using comments:

- Improved code clarity: Comments help explain what the code is doing, making it easier to understand and maintain.

- Documentation: Comments serve as documentation within your SQL scripts, providing information about the purpose of the code, specific logic, and limitations.

- Collaboration: By adding comments, you can communicate your intentions to other developers working on the same project, fostering better collaboration and understanding.

- Debugging: Comments can provide clues about the expected behavior of your code, aiding in debugging and troubleshooting issues.

Single-line comments: Start with -- and apply to the rest of the line.

```
SELECT * FROM users; -- Retrieves all users
```

Multi-line comments: Start with /* and end with */, allowing comments to span multiple lines.

```
/*This query provides the average
of each customer who has more than
five orders.*/

SELECT customer_id, AVG(order_total) AS avg_or-
der_value

FROM orders

/*Where customer_id = 23425

AND order_dt > '1JAN2023'    use when looking for
                              Specific examples*/

GROUP BY customer_id

HAVING COUNT(*) > 5;
```

Best practices for writing comments:

- Use clear and concise language. Avoid technical jargon and unnecessary details.
- Describe the purpose and functionality of the code. Explain what the code is doing and why it is needed.
- Document any assumptions or limitations. Highlight any potential issues or considerations.
- Update comments when the code changes. Keep your comments consistent with the actual code implementation.

By incorporating comments into your SQL code, you can significantly improve its readability, maintainability, and overall quality. This not only helps you understand and work with your code more effectively but also facilitates collaboration and ensures your intentions are clear to others.

Chapter 8: Virtual Tables

There may be a variety of virtual tables in any database. Virtual tables in SQL Oracle can be particularly helpful when writing queries This chapter will only discuss those virtual tables that are common to all databases.

A virtual table is a conceptual table that doesn't physically exist in the database but rather acts as a dynamic representation of data based on specific rules or algorithms. Unlike traditional tables stored on disk, virtual tables don't require dedicated storage space and are generated on demand when accessed.

There are numerous virtual tables that are common to most databases. This chapter will cover the most useful of those.

DUAL

The DUAL table in SQL is a special table that contains a single row and a single column. The column is called DUMMY and it always has the value X. It is often used as a placeholder in queries where a single row is needed.
You can use DUAL to check conditional logic, arithmetic calculations, date calculations, and other functions.

For example, if you are trying to get a query to return a specific date using the NEXT_DAY function, the following query could be used to ensure you will be returning the expected date.

```
SELECT (NEXT_DAY (TRUNC ((SYSDATE) +28), 'Saturday'))
FROM DUAL;
```

This will help you to see if the functionality is working correctly and that you are getting the results you expect. In the above example, if the system date were 12/10/2023 then the result would be 1/13/2024. The Saturday date after the system date would have been December 16th and 28 days later would be 1/13/2024.

```
SELECT TO_CHAR(SYSDATE, 'DD/MM/YYYY') FROM DUAL;
```

In this example, you can see how the TO_CHAR is producing the date in the format you desire.

```
SELECT DBMS_RANDOM.VALUE AS random_number FROM DUAL;
```

This query uses the DBMS_RANDOM.VALUE function to generate a random number between 0 and 1. DUAL enables us to easily obtain this system-generated value without any data manipulation.

These are just a few examples of how DUAL can be used for testing queries in Oracle SQL. Its versatility and simplicity make it a valuable tool for developers and database administrators alike.

ALL_TAB_COLUMNS

ALL_TAB_COLUMNS is a data dictionary table in Oracle SQL that provides a comprehensive overview and detailed information about all columns in all accessible tables, views, and clusters in the database. It is a one-stop shop for all the information in your database, enabling users to quickly identify and understand the structure of the underlying data.

It also helps users to potentially find columns that they know exist in the database but may not know where. Thus, it can be used as a reference point to figure out which tables and columns you may wish to include in the query and is populated with detailed information about each table and its corresponding columns, including:

- **Table_Name:** Name of the table, view, or cluster.
- **Column_Name:** Name of the column.
- **Data_Type:** Data type of the column (e.g., VARCHAR2, NUMBER).
- **Data_Length:** Length of the data type in bytes.
- **Nullable:** Indicates whether the column allows null values.
- **Data_Default:** Default value assigned to the column if no explicit value is provided.
- **Owner:** Owner or schema of the table, view, or cluster.

The following is an example of a query that will find tables or columns in the database that might be otherwise tough to locate in large database using the **ALL_TAB_COLUMNS** table.

```
SELECT a.owner, a.table_name, a.column_name

FROM all_tab_columns a

WHERE (a.table_name LIKE '%OCCUPATION%'
       OR a.column_name LIKE '%ONET%')

AND owner IN ('SCHEMA_1', 'SCHEMA_2', 'SCHEMA_4');
```

The results of this query would return any columns that have the characters onet anywhere in the name and any tables that have the characters occupation anywhere in the name. Results might look like the following:

OWNER	TABLE_NAME	COLUMN_NAME
SCHEMA_2	WORKER	MONETARY_AMT
SCHEMA_2	PRODUCTS	BALLONET
SCHEMA_4	OCU_CODE	ONET_CODE
SCHEMA_4	OCU_CODE	ONET_DESCRIPTION
SCHEMA_1	CUSTOMERS	ONETIME_CUSTMR

In the above query the user was looking for a table that contained Occupational or ONET codes. As we can see from the results the table he needs to query against is the OCU_CODE table, which is found in SCHEMA_4. As we can see from the results, there were no tables in the database that contained the characters "occupation" as part of their name.

VIEWS

In Oracle SQL, a VIEW offers a unique way to manage and access data. A VIEW acts as a virtual table, representing the result set of a stored query without physically storing any data itself. This flexible approach allows for tailoring data access based on specific requirements.

VIEWs can be used like tables in that you can query against them and join other tables to them. (See more information on Views in Chapter 9).

ALL_VIEWS

ALL_VIEWS is a data dictionary table in Oracle SQL that provides a comprehensive overview and detailed information about all views in your database.

This table is very similar to the ALL_TAB_COLUMS table, however instead of showing you all tables, it will show you all VIEWS that have been created for your database with their associated schemas, as well as the query that was used to create the view.

Below is an example of a query that might be used to query the ALL_VIEWS table.

```
SELECT *

FROM all_views

WHERE owner = 'SCHEMA_1';
```

Chapter 9: Views

A view in an Oracle database is a virtual table that presents data retrieved from one or more underlying tables based on a defined query. It acts like a pre-written SQL query stored in the database, but unlike a table, it does not physically store any data itself.

Through filtering, aggregation, joining, and manipulation, views can present a specific subset or transformation of data from various sources. This simplifies complex queries by hiding the underlying data structures and logic, making it easier for users to work with information without needing to understand intricate details.

Furthermore, VIEWs promote code reusability. Instead of writing complex queries repeatedly, you can define a view to represent the desired data transformation and use it throughout your applications, enhancing development efficiency.

VIEWs also play a crucial role in data security. By granting users permission to access specific views while restricting access to the underlying data, you can control and monitor information exposure, ensuring data privacy and integrity.

In some cases, using views can improve query performance. Materialized views pre-compute the result set of the underlying query, potentially offering faster response times for frequently accessed data. This can be beneficial for performance-critical applications.

Overall, Oracle SQL views offer a flexible tool for data management. They simplify data access, promote code reusability, and enhance data security, making them a valuable asset for developers and database administrators seeking efficient and effective data management solutions.

Creating Views

In SQL, a view is a virtual table that provides a customized representation of the underlying data stored in one or more tables. It acts as a filter that allows users to see and interact with data in a specific way, without modifying the actual data itself.

Views can simplify complex queries by pre-defining frequently used joins, conditions, and aggregations. This makes it easier to write and maintain queries that involve multiple tables or complex filters. They can be used to restrict access to sensitive data by only exposing specific columns or rows. This helps ensure data security and prevents unauthorized users from seeing or modifying critical information.

Views can also improve query performance by pre-calculating and storing frequently accessed data. This can be beneficial for queries that involve large datasets or complex calculations. Additionally, views can be used to present data in a specific format or order, making it easier for users to understand and analyze the information.

Consider creating a view if you find yourself repeatedly writing the same complex query, need to restrict data access for certain users, want to improve the performance of frequently used queries involving large datasets or complex calculations, or desire to standardize data presentation across your application for consistency.

By leveraging views, you can streamline your code, enhance data security, optimize performance, and ensure consistent data presentation, ultimately improving the efficiency and effectiveness of your SQL operations.

Use the CREATE VIEW statement to create the view.

1. The CREATE VIEW statement takes two arguments: the name of the view and the SELECT statement that defines the view.

2. The SELECT statement can be used to select data from one or more tables.

3. The CREATE VIEW statement also allows you to specify the following options:

    a. DISTINCT: This option tells the database to return only distinct rows from the view.

    b. WITH CHECK OPTION: This option tells the database to only allow changes to the view that are valid for the underlying tables.

Once you have created the view, you can use it like any other table.

Example:

```
CREATE VIEW my_view AS

SELECT

   name

 , email

FROM users;
```

This view will create a view called **my_view** that contains the **name** and **email** columns from the users table.

To use the view, you can use it in a **SELECT** statement like this:

```
SELECT *

FROM my_view;
```

Chapter 10: Creating and Populating Tables

In this chapter, we'll explore the fundamentals of table creation and population in Oracle SQL. From defining columns and data types to inserting initial values, we'll discuss the essential techniques for structuring and storing your data.

When making changes to a database the changes are temporary until the "COMMIT;" command is run. COMMIT will commit or save your changes permanently to the database, otherwise the changes you make will be undone when you log out of your querying tool.

Creating a Table Using Predefined Column Names and Types

You can create a table with predefined column names and column types then insert data into that table.

The following is the syntax to do this:

```
CREATE TABLE schema_name.table_name (

column_name data_type [CONSTRAINT constraint_name]
[DEFAULT default_value] [NULL | NOT NULL],

column_name data_type [CONSTRAINT constraint_name]
[DEFAULT default_value] [NULL | NOT NULL],

... );

COMMIT;
```

- **Table_name** is the name of the table you want to create.
- **Column_name** is the name of the column you want to create.
- **Data_type** is the data type of the column you want to create.
- **Constraint** constraint_name is the name of a constraint you want to apply to the column.
- **Default** default_value is the default value for the column.
- **NULL / NOT NULL** specifies whether the column can contain NULL values.

Example:

```
CREATE TABLE products_table
(
  PRIMARY_KEY     NUMBER(18)        NOT NULL,
  CUSTOMER_ID     NUMBER(18)        NOT NULL,
  ORDER_ID        NUMBER(10),
  ORDER_DT        DATE              NOT NULL,
  SURVEY_CD       VARCHAR2(2 BYTE)  NOT NULL,
  UPSALE_CD       VARCHAR2(2 BYTE)  NOT NULL,
  UPDATE_IND      NUMBER
);
      COMMIT;
```

Creating a Table Using a Query

You can create a Table_B By specifying its structure and data using the **CREATE TABLE** command followed by a query. Remember to commit your changes.

The following is the syntax to do this:

```
CREATE TABLE schema_name.table_name AS (query);

COMMIT;
```

Example:

```
CREATE TABLE sales_schema.products_table  AS

(SELECT cl.P_KEY4, cl.P_KEY2, pt.P_KEY1, cl.eff_dt,

  FROM SCHEMA_1.TABLE_1 cl, SCHEMA_1.TABLE_4 pt

 WHERE cl.DATE BETWEEN '1-JAN-2017' AND '31-DEC-2017'

   AND cl.CODE IN ('B', 'L', 'T')

   AND cl.P_KEY2 = pt.P_KEY2);

COMMIT;
```

Populate a Table with Data

Use **INSERT INTO** to manually insert data into a row.

The following is the syntax to do this:

```
INSERT INTO schema_name.table_name (column1, column2,
...)

VALUES (value1, value2, ...);

COMMIT;
```

Example:

```
INSERT INTO sales_schema.products_table
        (SALES_ID, CUSTOMER_ID, STORE_ID
       , PURCHASE_DT, STATE, REPEAT_IND, VIP_IND)

VALUES  ('10716', '478486', '0011455714'

       , to_date('02/04/2017' , 'mm/dd/yyyy')

       , 'UT', '1', '0');

COMMIT;
```

To populate a table with data <u>using a query</u> use the following syntax:

```
INSERT INTO table_name (column1, column2, ...)

SELECT column1, column2, ...

FROM source_table;

COMMIT;
```

Example:

```
INSERT INTO sales_schema.products_table

(P_KEY4, P_KEY2 , P_KEY1, eff_dt, state_cd, bkdt_cd,
update_ind)

SELECT cl.P_KEY4, cl.P_KEY2 , pt.P_KEY1
     , cl.eff_dt, cl.state_cd
     , cl.bkdt_cd,'0' update_ind

FROM SCHEMA_1.TABLE_1 cl, SCHEMA_1.TABLE_4 pt

WHERE cl.DATE BETWEEN '1-JAN-2017' AND '31-DEC-2017'

       AND cl.P_KEY2  = pt.P_KEY2;

     COMMIT;
```

Using MERGE to Insert Data

The **MERGE** command is a statement in SQL that combines the functionality of **INSERT**, **UPDATE**, and **DELETE** statements into a single operation. This makes it a versatile tool for efficiently manipulating data in your database.

You can use the **MERGE** statement to insert additional information into a table from a query, synchronize data between two tables, update existing data based on new information, insert new data that meets specific criteria, and perform conditional deletes based on matched rows.

By understanding the **MERGE** command and its applications, you can effectively manipulate data in your SQL queries and achieve better performance and code clarity.

```
MERGE INTO schema.table_name d

USING (query)q

ON (d.P_KEY4 = q.P_KEY4) --to tie query to table

WHEN MATCHED THEN UPDATE SET

 d.dur = wk.dur
--tie the columns in the query to columns in the ta-
ble

,d.earn = wk.earn

,d.mod_earn = wk.mod_earn;

 COMMIT;
```

Another example of **MERGE:**

```
MERGE

INTO customers

USING new_customers

ON customers.customer_id = new_customers.customer_id

WHEN MATCHED THEN

    UPDATE SET email = new_customers.email,

               phone = new_customers.phone

WHEN NOT MATCHED THEN

    INSERT (customer_id, name, email, phone)

    VALUES  (new_customers.customer_id

           , new_customers.name, new_customers.email

           , new_customers.phone);
```

This **MERGE** statement updates the email and phone number of existing customers based on data from the new_customers table. It also inserts any new customers not present in the existing table.

Updating data in a Table

To update data in single column, use the **UPDATE** statement. **UPDATE** is used to insert data into an existing table.

The following is the syntax to do this:

```
UPDATE schema.table_name
   SET column_name = new_value
 WHERE condition;
COMMIT;
```

The WHERE clause is in the UPDATE statement is optional but almost always used in real world usage.

Example:

```
UPDATE schema.table_name
   SET update_ind = 1
 WHERE eff_dt > '1Jun2017';
COMMIT;
```

This will insert a "1" in every row of the update_ind column where the eff_dt is greater than 1 June 2017.

Remove All Rows from a Table

Use **TRUNCATE TABLE** to remove or delete all rows from a table. The table will still exist, but it will be void of data.

```
TRUNCATE TABLE schema.table_name;

COMMIT;
```

Delete a Table

You can delete an entire table using **DROP TABLE**. After executing this command the table ceases to exist.

```
DROP TABLE schema.table_name;

COMMIT;
```

Delete Specific Columns from a Table

Remove specific columns from a table using **ALTER TABLE**.

```
ALTER TABLE schema.table_name DROP (column1,column2);

COMMIT;
```

Rename Columns

Rename columns using **ALTER TABLE**.

```
ALTER TABLE schema.table_name RENAME COLUMN column1
TO column2;

COMMIT;
```

Chapter 11: Joining Tables

Table joins are fundamental operations in relational databases, allowing you to combine data from multiple tables based on specific relationships between them. They play a crucial role in data analysis, reporting, and application development, enabling you to extract meaningful insights and generate comprehensive information from your database.

Table joins enable you to discover relationships and patterns within your data that would be hidden if the tables were kept isolated. By joining tables on common attributes, you can identify correlations, trends, and associations that can lead to informed decisions and strategic planning. They are essential for data analysis and reporting as they allow you to extract specific information from multiple tables. This enables you to track trends, identify anomalies, and gain insights into various aspects of your business or organization.

The ANSI Join

Joining in the **FROM** clause (a more traditional way of joining tables).

1. INNER JOINS

The **INNER JOIN** combines data from two tables where there is a match on the joining column(s) in both tables. Remember, the **INNER** keyword is optional. This is the most common type of join.

Figure 7: Inner Join

```
SELECT A.customer_id, A.zip_code, B.or-
der_dt

FROM Customers A

INNER JOIN Orders B

ON A.Customer_id = B.Customer_id;
```

CUSTOMER_ID	ZIP_CODE	ORDER_DT
15640	84864	07 NOV 2020
15640	84864	15 JAN 2020
15642	74320	11 JUL 2020
15643	64321	08 MAY 2020

These data points are from records that link directly from the orders table to the customers table in a one-to-many relationship. Which is shown by the same customer_id displaying in the first and second rows. One customer can have many orders.

To add more tables, add the join type, list the table, and then the link as shown here:

```
SELECT Columns
  FROM Customers A
  INNER JOIN Orders B
  ON A.Customer_id = B.Customer_id
  INNER JOIN Inventory C
  ON B.Inventory_id = C.Inventory_id;
```

When writing queries with multiple tables, every table must be linked to another table in the query. There should be no break in the chain of tables. Every time you create a join ensure you are considering the relationship between the tables you are joining (whether it is a one-to-one, one-to-many, or a many-to-many relationship).

2. OUTER JOINS

The **OUTTER JOIN** brings in all rows from the indicated table and only the associated rows from the other table or nulls if a matching row doesn't exist. This will result in null values in the records of the first table where there are no records to link on in the second table. Remember, the **OUTER** keyword is optional. This is the second most common type of join.

The following depicts the left outer join:

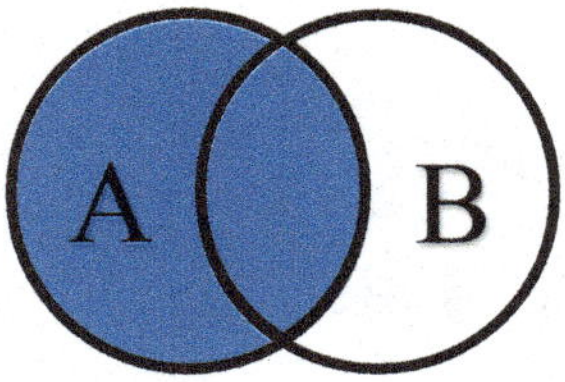

Figure 8: Left Outer Join

```
SELECT A.student_name, B.class

FROM Students A LEFT OUTER JOIN Classes B

ON A.Student_ID = B.Student_ID;
```

The results from this query might look like the following:

STUDENT_NAME	CLASS
Johnny Beebe	Math 1010
Michelle Cash	Science 2400
Kim Badame	
Jolynn Kamando	History 3400

In this example, Kim Badame has a NULL in the CLASS column because she does not have a row in the class table, but does have a row in the students table.

The following depicts the right outer join:

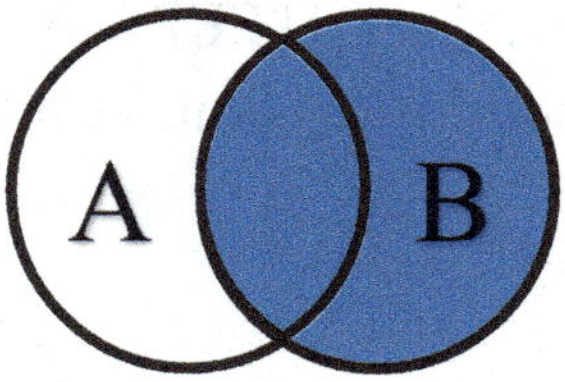

Figure 9: Right Outer Join

```
SELECT Columns

FROM Customers A

RIGHT OUTER JOIN Orders B

ON A.Customer_ID = B.Customer_ID;
```

Adding a null filter changes the nature of the **OUTER JOIN**

The following depicts the left outer join with a null filter:

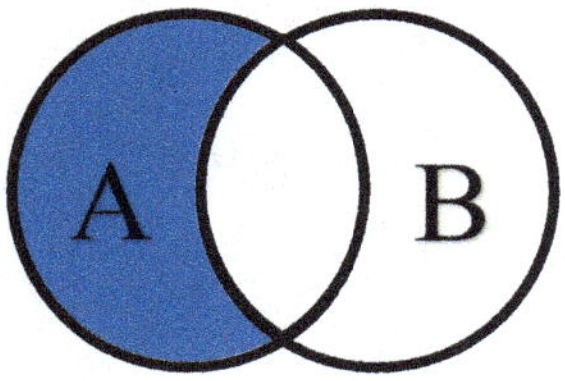

Figure 10: Left Outer Join Null B.key

```
SELECT Columns

FROM Students A

LEFT OUTER JOIN Classes B

ON A.Student_ID = B.Student_ID

WHERE B.Student_ID IS NULL;
```

The following depicts the right outer join with a null filter:

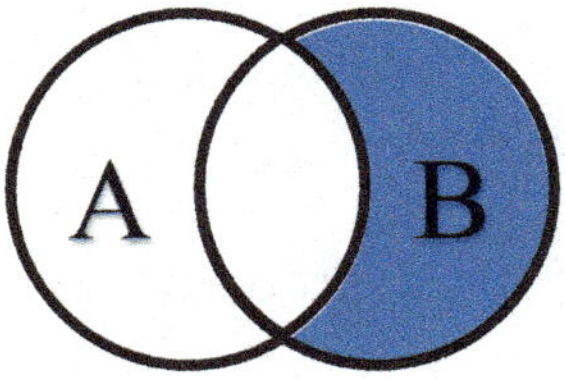

Figure 11: Right Outer Join Null A.key

```
SELECT Columns

FROM Customers A

RIGHT OUTER JOIN Orders B

ON A.Customer_ID = B.Customer_ID

WHERE A.Customer_ID IS NULL;
```

The **FULL OUTTER JOIN** will return every record from every table. It collects every single record from both tables involved, ensuring a comprehensive overview, even if some information is missing in either one. If there is a conventional match, it is made. If either side has missing data, it is replaced by nulls.

Figure 12: Full Outer Join

```
SELECT Columns
FROM Customers A
FULL OUTER JOIN Orders B
ON A.Customer_ID = B.Customer_ID;
```

The following depicts the full outer join with a null filter:

Figure 13: Full Outer Join Nulls excluded

```
SELECT Columns
FROM Customers A
FULL OUTER JOIN Orders B
ON A.Customer_ID = B.Customer_ID
WHERE (A.Customer_ID IS NULL
        OR B.Customer_ID IS NULL);
```

3. CROSS JOINS

The **CROSS JOIN** is the deliberate creation of a cartesian product from two tables. There are no join columns specified, so every possible combination of rows between the two tables is produced. This is virtually never a desirable join to use.

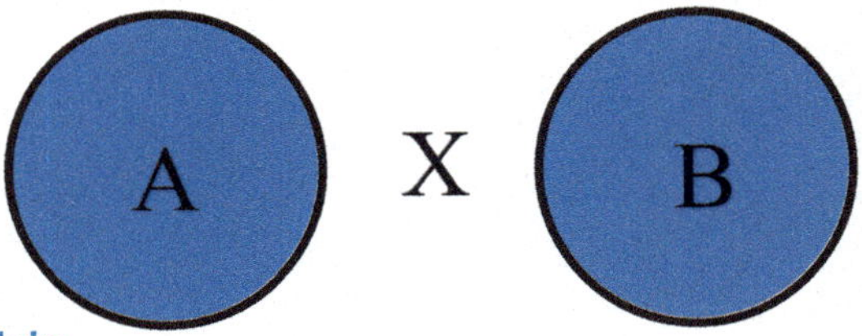

Figure 14: Cross Join

```
SELECT Columns
FROM Customers A
CROSS JOIN Orders B;
```

4. NATURAL JOINS

A **NATURAL JOIN** is a variant on an **INNER JOIN**. The join columns are determined implicitly, based on the column names. Columns with the same name between the two tables are assumed to be join columns.

- Using a **NATURAL JOIN** is discouraged because if someone adds a new column to one of the tables with the same name as a column in the other table, it may break existing natural joins. It's effectively a potential issue waiting to happen.

- You can't apply any aliased filters to columns used in natural joins because **NATURAL JOIN** implicitly determines the join columns based on the column names. If you try to use an aliased filter with a column that is used in a **NATURAL JOIN**, the filter will not be applied.

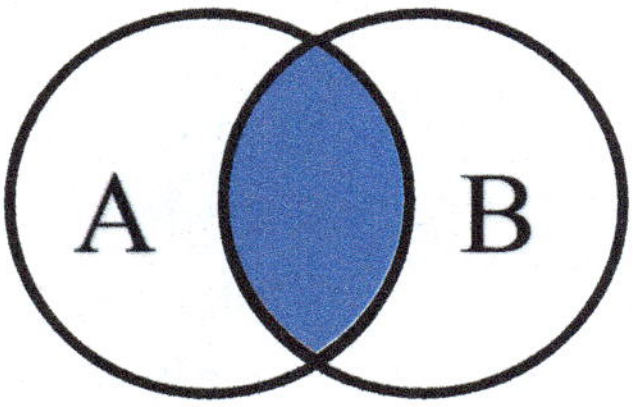

Figure 15: Natural Join

```
SELECT Columns

FROM Customers A

NATURAL JOIN Orders B;
```

The **INNER JOIN ... USING** is almost a half-way house between a conventional **INNER JOIN** and a **NATURAL JOIN**. The join is made using columns with matching names in each table, but you have to specify the columns to be used, not the whole condition. This allows you to join on a subset of the columns common to both tables. Similar to the **NATURAL JOIN**, you can't apply any aliased filters to columns used in the join, but if you remove the alias it works.

```
SELECT Columns

FROM Customers a

JOIN Orders b USING (Customer_ID);
```

The Oracle Join

ANSI joins are standardized and supported by all major RDBMS, while Oracle joins are proprietary to Oracle Database. The ANSI join is done exclusively in the **FROM** clause whereas the Oracle join is exclusively in the **WHERE** clause (not traditional).

1. INNER JOINS

The **INNER JOIN** in the WHERE clause uses the = sign between the two keys. (e.g., a.key = b.key)

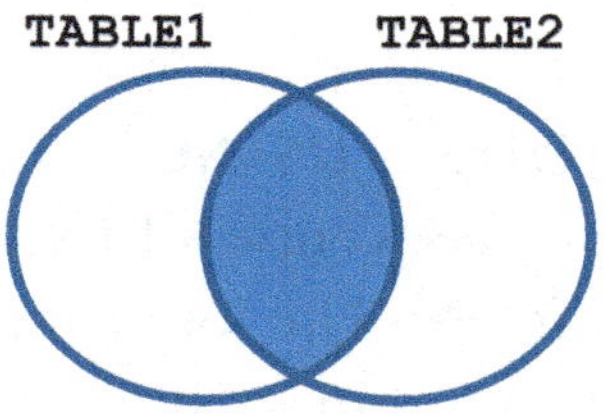

Figure 16: Inner Join

```
SELECT Columns
FROM Students AS A, Classes AS B
WHERE A.Student_ID = B.Student_ID;
```

In this example you are asking to see only those students with classes assigned to them.

2. OUTER JOINS

An outer join is done by placing a (+) on the **WHERE** clause after the column where NULLs are expected. For example, a **LEFT OUTER JOIN** can be written as:

```
SELECT a.student_name, b.class

FROM Students AS A, Classes AS B

WHERE A.Student_ID = B.Student_ID(+);
```

You are expecting students with null classes so the (+) goes next to the classes column in the join because that is where the nulls are expected.

This query is saying you want to see all the students and only those classes associated with students, but if there are students without classes you still want them to show up in the results. So, there may be students enrolled in the school that have not yet signed up for any classes.

The above query might return the following results:

STUDENT_NAME	CLASS
Bill Godfrey	Math 1010
Julie Nigh	Science 2400
William Washington	
Vanessa Cody	History 3400

With an inner join William Washington would have been excluded from the results because he has no associated row/student_id in the class table.

A **RIGHT OUTER JOIN** can be written by putting the symbol on the other side of the join:

```
SELECT Columns

FROM Customers AS A, Orders AS B

WHERE A.Customer_ID (+) = B.Customer_ID;
```

To convert your query from ANSI to ORACLE or from ORACLE to ANSI use the following website:

https://www.jooq.org/translate/

Chapter 12: Combining Queries

The following functions are placed between two queries. To use any of the following functions the queries being combined must have the same column types with the same number of columns.

If the first column in the first query is a date, then the first column in the other query or queries must also be dates, if the second column is a number in the first query then all other queries you want to combine must also have numbers in the second column. The same is true for the third, fourth, and any other columns you would want to use.

1. UNION

A union statement is a tool in SQL that allows you to combine the results of two or more **SELECT** statements into a single result set. It is particularly useful for merging data from multiple sources or retrieving data that meets different criteria.

UNION removes duplicate records effectively like a **SELECT DISTINCT** (where all columns in the results are the same). Corresponding columns must have compatible data types.

In essence, a union statement acts as a data aggregator, merging information from multiple sources into a single, unified result set, making it an invaluable tool for data analysis and reporting.

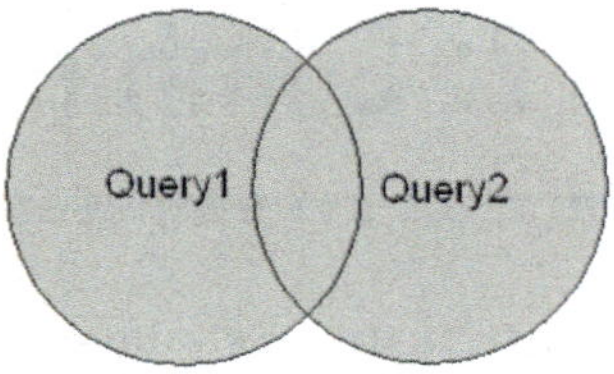

Figure 17: Union

```
SELECT A.NAME, B.DATE
  FROM Table_A AS A
          UNION
  SELECT C.NAME, C.DATE
  FROM Table_C AS C;
```

The data from this query will return the following:

NAME	*DATE*	
Will Wilson	*1 Mar 2022*	
Lana Smith	*6 Apr 2022*	*Results from Table_A*
Fred Mutz	*9 Feb 2002*	
Michael F	*5 May 2022*	*Result from both tables combined into one row*
Justin	*2 Sep 2022*	
Kevin	*7 Nov 2022*	*Results from Table_C*
Casey	*2 Jul 2022*	

2. UNION ALL

Will not eliminate duplicate rows, instead it just pulls all rows from all tables fitting your query specifics and combines them into a table (where all columns in the results are the same). Corresponding columns must have compatible data types.

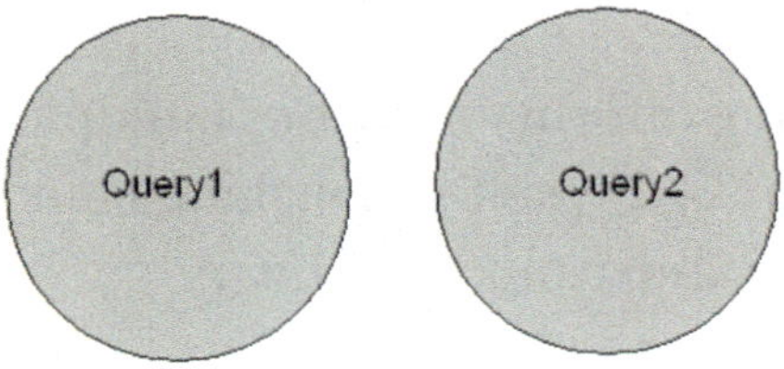

Figure 18: Union All

```
SELECT A.NAME, B.DATE
FROM Table_A A
        UNION ALL
SELECT C.NAME, C.DATE
FROM Table_C C;
```

The data from this query will return looking like the following:

NAME	DATE	
Will Wilson	1 Mar 2022	
Lana Smith	6 Apr 2022	Results from Table_A
Fred Mutz	9 Feb 2002	
Michael F	5 May 2022	Michael F is a duplicate row
Michael F	5 May 2022	
Justin	2 Sep 2022	
Kevin	7 Nov 2022	Results from Table_C
Casey	2 Jul 2022	

3. MINUS

Returns the difference between the first and second query. Subtracts the rows in one query from the rows in another query. MINUS Returns only those results that are in the first query and not matched with rows in the second (where all columns in the results are the same). Corresponding columns must have compatible data types.

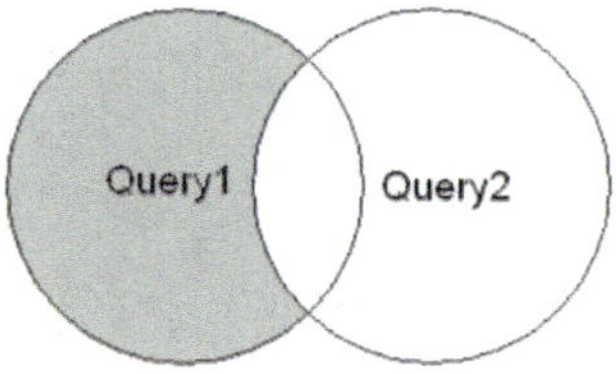

Figure 19: Minus

For example, if you wanted to see the rows that were being excluded from a query where you did not truncate a column that had a date and time stamp, you could use minus to see all the rows you are not capturing because you did not truncate the date.

```
SELECT c.key ,c.date

FROM SCHEMA1.table1 c

WHERE TRUNC(c.date) BETWEEN '2Mar2018' AND '3Mar2018'

MINUS

SELECT c.key ,c.date

FROM SCHEMA1.table1 c

WHERE c.date BETWEEN '2Mar2018' AND '3Mar2018';
```

The resulting rows from this query would be all of the rows from the first query that are not showing up in the second query, (which would be all the rows with a date of 3 Mar 2018.)

Generally, if you made any change to your query and were expecting to see the same number of rows return after the change, but instead saw an unexpected number of rows returned, you could see which rows were added or taken away by using **MINUS**.

For example, if you lost rows you could take the query before you changed it add a **MINUS** underneath it then under that add the same query with the change you made. The result would be the rows that were lost in the change. Then you can determine if you need to write the query differently if those were rows you decided you needed to keep.

INTERSECT

Returns the rows that are found in both the first, and second query (where all columns in the results are the same). Corresponding columns must have compatible data types

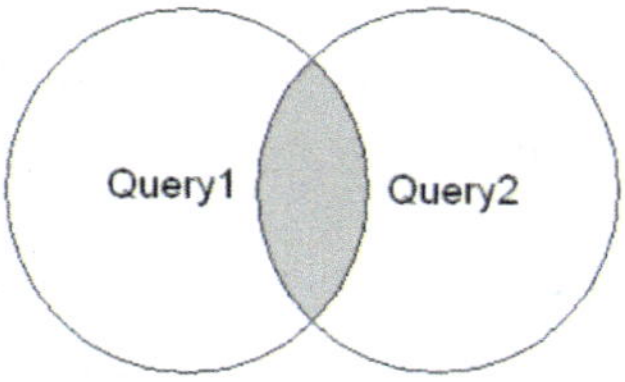

Figure 20: Intersect

Chapter 13: Subqueries and WITH

Subqueries are SQL queries that are nested within other SQL queries. They can be used to perform a variety of tasks, such as:

- Filtering data based on the results of another query
- Performing calculations on data from another query
- Joining data from multiple tables

Subqueries can be very useful for simplifying complex queries and making them more readable and maintainable. They can also be used to improve the performance of queries by avoiding unnecessary joins.

To create a subquery place parenthesis at the beginning and end of the query then place it within another query to make it a subquery.

Using Subqueries in the SELECT statement

You can use subqueries in the **SELECT** statement to perform conditional operations.

```
SELECT CASE WHEN c.P_KEY4 IN (
                          SELECT t.P_KEY2
                             FROM SCHEMA_1.TABLE_1 t
                            WHERE mod_cd = 'PI'
                               )
             THEN 1 ELSE NULL END UI_IND
FROM SCHEMA_1.TABLE_2 c;
```

Using Subqueries in the FROM statement

Subqueries in the **FROM** statement can be used to turn your query into a table from which you can pull data.

```
SELECT c.*, cl.limit
FROM SCHEMA_2.CUSTOMERS c
  , (SELECT cc.customer_id, cc.limit
      FROM SCHEMA_1.CREDIT_CARD cc
       WHERE cc.limit > 10000) cl
WHERE c.customer_id = cl.customer_id
AND c.date > '1JAN2020';
```

In this example, because the subquery is in the **FROM** statement, it becomes a table that can be queried against. It is given the alias "cl" and linked to the CUSTOMERS table using the customer_id which is in both the CUSTOMERS table and the subquery.

Using Subqueries in the WHERE statement

Subqueries in the **WHERE** statement are used to filter rows based on conditions defined in the subquery.

```
SELECT *
FROM SCHEMA_2.CUSTOMERS c
WHERE c.eff_dt> '1jan2018'
AND c.customer_id IN(SELECT cc.customer_id
                       FROM SCHEMA_1.CREDIT_CARD cc
                      WHERE cc.limit > 10000);
```

In this example the subquery is creating a specific subset of customers. In the outer query we are limiting the results to only the rows that have the same customer_id as exists in the subquery. Thus, all results are returning from customers who are identified as having a high credit limit, of more than $10,000 since January 2018.

The "WITH" Clause

The **WITH** statement allows you to create a common table expression (CTE) that can be referenced within your main query. It helps simplify complex queries and improve readability.

The syntax for the **WITH** statement is as follows:

```
WITH (insert alias name)

  AS (query)

SELECT (columns from query)

FORM (alias name used after WITH)

WHERE (insert any conditions);
```

Example:

```
WITH Contact AS (
          SELECT  c.P_KEY1
               , c.UNIQUE_ID
               , c.OTHER_ID
               , c.last_name
               , a.addr1 AS street
               , a.city
               , a.state
               , a.zip
          FROM SCHEMA_3.TABLE_A1 c
             , SCHEMA_3.TABLE_A2 u
             , SCHEMA_1.TABLE_4 a
          WHERE c.P_KEY6 = u.P_KEY6
          AND c.P_KEY1 = a.P_KEY1
          AND u.type_cd = 'S'
             )
SELECT *
FROM contact
WHERE city = 'JUNCTION';
```

In this example **WITH** is giving the subquery an alias "Contact" the query just under the subquery is selecting information from "Contact" and adding the necessary filters.

Chapter 14: Common Operators

The most basic way to filter data is using comparison operators. These operators and their uses may be worth committing to memory.

Type	Operator	Description	Example
Comparison	=	Equal to	b.FIRST_NAME = 'BOB'
	<> or !=	Not equal to	b.DEPARTMENT <> 'SALES'
	>	Greater than	CREATED_DT > '1 JUL 2020'
	<	Less than	CREATED_DT < '1 JUL 2020'
	>=	Greater or equal to	CREATED_DT >= '1 JUL 2020'
	<=	Less than or equal to	CREATED_DT <= '1 JUL 2020'
Arithmetic	+	Add numbers of two columns together (or plug in your own number)	c.SALES_AMT + p.TAX_AMT
	-	Subtract numbers of two columns (or plug in your own number)	c.GRANT_AMT - c.TOTAL
	/	Divide numbers in two columns (or plug in your own number)	c.SALES_AMT / 26

Type	Operator	Description	Example
Arithmetic	*	Multiply numbers in two columns (or plug in your own number)	`c.SALES_AMT * 26`

	^	Exponentiation	`SELECT 5 ^ 2`
			`FROM Dual;`
			
			Output = 25

	%	When used in a LIKE statement the % becomes a wildcard. It represents zero, one, or more characters of any kind.	Place % within the pattern to match any substring of the same length. For example, `LIKE 'a%e'` will match `"apple", "ape", "plane",` etc.
Logical	**BETWEEN**	BETWEEN is an inclusive range	`CREATED_DT BETWEEN ‘1 JUL 2020’ AND ’31 JUL 2020’`
	LIKE or **NOT LIKE** (Use the wildcard character “%”)	Finds all records that begin with characters before the wildcard ‘%’	`a.FIRST_NAME LIKE ‘TACO%’`
		Finds all records that end with characters after the wildcard ‘%’	`a.FIRST_NAME LIKE ‘%TACO’`

Type	Operator	Description	Example
Logical	**LIKE** or **NOT LIKE** (Use the wildcard character "%")	Finds all records that contains the characters between the wildcards '%' that exist somewhere in the text string.	`a.FIRST_NAME LIKE '%TACO%'`
		Finds all records that begin and end with characters that are before and after the wildcard '%'	`a.FIRST_NAME LIKE 'TA%CO'`
	(use the place holder character "_")	The 2nd character must be "A" and "CO" must be at the end of the string.	`a.FIRST_NAME LIKE '_A%CO'`
	AND	The AND keyword is used for additional filters after the first WHERE filter.	`WHERE P_IND = 5 AND sales_amt > 100`
	(condition **OR** condition)	Used after WHERE or AND filter. Will return the values if either argument is true (parenthesis are required for this to work correctly).	`(sales_ind = '1' OR refered_ind = 'R')`

<table>
<tr><th>Type</th><th>Operator</th><th>Description</th><th>Example</th></tr>
<tr><td rowspan="6">Logical</td><td>NOT</td><td>Negates the condition.</td><td>(See page 92)</td></tr>
<tr><td>IN
or
NOT IN</td><td>This is like = or <> but with multiple values or find a list of precise values.</td><td>WHERE d.SALES_CD IN ('P', 'U', 'L', 'N')
AND d.STATE_CD NOT IN ('LA', 'NV', 'IL')</td></tr>
<tr><td>IS NOT NULL or IS NULL</td><td>Finds rows where there is or is not a NULL.</td><td>WHERE b.address IS NOT NULL</td></tr>
<tr><td>IS TRUE
or
IS NOT TRUE</td><td>Boolean truth value test.</td><td>WHERE p.validation IS TRUE</td></tr>
<tr><td>IS FALSE
or
IS NOT FALSE</td><td>Boolean truth value test.</td><td>WHERE p.sales_made IS FALSE</td></tr>
<tr><td>IS NOT DISTINCT FROM</td><td>Is equal to value or both are nulls (missing data).</td><td>a.debt IS NOT DISTINCT FROM b.receivables</td></tr>
<tr><td rowspan="2">Other</td><td>||</td><td>Concatenate - Links two columns into one.</td><td>eg. (h.first_name|| h.last_name) the result would look like “JohnDoe”

To add a space in between the columns add a space wrapped in double quotation marks with an additional concatenation as follows: (h.first_name||' '|| h.last_name) the result would look like “John Doe”</td></tr>
<tr><td>-- or /* */</td><td>Comment out</td><td>--words or /* words, words*/</td></tr>
</table>

Chapter 15: Common Functions

Like operators, functions take the data as input and produce equivalent output manipulating the data to achieve the desired outcome. They are like specialized tools you can use to perform specific tasks, while operators are like everyday tools available for common, basic tasks. With functions, you can take multiple arguments at once allowing for more complex operations.

ADD_MONTHS()

Adds a specified number of months to your date.

```
ADD_MONTHS(column, 3)
```

This will add three months to your date.

ADD_MONTHS can calculate the due date based on a start date and number of months.

```
SELECT ADD_MONTHS(TO_DATE('2023-10-26', 'YYYY-MM-DD'), 6) AS target_date

FROM DUAL;
```

This query adds 6 months to the date "2023-10-26" and returns the result in the "target_date" column.

ADD_MONTHS can also find dates within a specific month range.

```
SELECT * FROM events

WHERE e.event_date BETWEEN ADD_MONTHS(SYSDATE, -3)

AND SYSDATE;
```

This query selects all events that occur within the past 3 months, including the current month.

ADD_MONTHS will also calculate the due date based on a start date and number of months:

```
SELECT order_date

, ADD_MONTHS(order_date, payment_terms) AS due_date

FROM orders;
```

This query retrieves the order date and payment terms for each order and then calculates the due date by adding the specified number of months (in the payment terms) to the order date.

AS

Used to change a column name (assign alias) when viewing results.

```
SELECT b.employee AS worker
```

One can also use a space instead of AS

```
SELECT b.employee worker
```

AVG ( )

The **AVG()** function in SQL returns the average value of a numeric expression.

```
AVG(column)
```

The *column* can be any expression that returns a numeric value. For example, the following are valid expressions for the **AVG()** function:

```
SELECT w.department_name, AVG(w.salary)
FROM schema1.workers w
GROUP BY w.department_name;
```

This query will provide a list of departments with the average salary being earned in each department.

CAST()

This function can be used to convert a string, or a date into another data type. It can also be used to convert NULL values to other data types.

```
CAST(column AS data_type(length))
```

Note: (length is optional)

For example:

```
SELECT CAST(created_date AS TIMESTAMP) FROM sales;
SELECT CAST(date_loaded AS DATETIME) FROM sales;
SELECT CAST(set_ind AS VARCHAR(10)) FROM sales;
SELECT CAST([row] AS INT) FROM sales;
```

COALESCE()

Return the first non-NULL value from the columns listed.

```
COALESCE(c.UNIQUE_ID ,c.OTHER_ID)
```

For example, the following query will capture the SSN when it is available. When it is not, it will capture the Customer_ID.

```
SELECT COALESCE(w.SSN ,w.CUSTOMER_ID) AS unique_id
FROM schema1.workers w;
```

COUNT()

Will count anything that has values if you use a column name. **COUNT(*)** counts every row in the data set.

```
COUNT(column)
```

For example:

```
SELECT COUNT(*)
FROM schema1.receivables;
```

You can count the values in a column by including the column name. In this case NULLs will not be counted.

```
SELECT COUNT(r.discount_amt)
FROM schema1.receivables r;
```

Note: **GROUP BY** is not needed if the aggregate function is by itself (no other columns).

DECODE()

DECODE is a function that is used to provide if-then-else type of conditional logic to SQL. This is much like the CASE function which is more commonly use and described in the next chapter.

```
DECODE (expression, search1, result1, search2,
       result2, ..., default)
```

Expression is the expression that will be evaluated.

Search1, search2, etc. are the search values that will be compared to the expression value.

Result1, result2, etc. are the results that will be returned if the expression value matches the corresponding search value.

Default is the result that will be returned if the expression value does not match any of the search values.

```
SELECT s.customer_id

, DECODE(s.sales_amt, 0, 0, s.paid_amt / r.sales_amt)

FROM sales s;
```

Means:(If column labeled SALES_AMT = 0 then 0 otherwise use this formula.)

When you get a "cannot divide by zero" error use the following:

```
DECODE (a.column,0,NULL)/ b.column
```

DISTINCT

Inside a table, a column often contains many duplicate values; and sometimes you only want to list the unique or distinct values.

```
SELECT DISTINCT column1, column2, …
FROM table_name;
```

DISTINCT can also function within a subquery or **WHERE** clause to filter data based on unique values in a specific column. This allows you to select only rows where the specified column contains distinct values.

For example, the following query retrieves customers who have placed at least one order:

```
SELECT *
FROM customers c
WHERE c.customer_id IN (
       SELECT DISTINCT o.customer_id
         FROM orders o );
```

EXISTS() or NOT EXISTS()

The **EXISTS** operator tests for the existence of any record in a subquery. It verifies the existence of data based on specified conditions within a subquery and limits your data to the matching records. A **NOT EXISTS** function excludes any records that match on the join.

When you use an **EXISTS** function, the related subquery should tie into the main query.

See the following example:

```
SELECT *
FROM customers c
WHERE EXISTS (
            SELECT 1
            FROM orders o
            WHERE o.customer_id = c.customers_id
            AND o.product_id = 1234
               );
```

This query selects all customers who have placed an order for the product with ID 1234. Instead of selecting orders directly, it uses **EXISTS** to ensure that at least one order exists for the specific product and customer.

In the subquery, the character "1" is used in the **SELECT** statement. This is just a manually input value that will return if the query returns something. It could be any number or letter on the keyboard since you are just checking for the existence of a row. For example, you could have put an 'x' instead of a 1 and gotten the same result.

EXTRACT()

The **EXTRACT** function is used for extracting specific date or time components from a datetime value. It allows you to isolate and analyze individual parts of a date and time, such as the year, month, day, hour, minute, or even millisecond.

```
SELECT EXTRACT(MONTH FROM TRUNC(created_dt)) = '5';

SELECT EXTRACT(DAY FROM TRUNC(created_dt)) = '15';

SELECT EXTRACT(YEAR FROM TRUNC(created_dt)) = '2022';
```

FLOOR() and CEIL()

The **FLOOR** function in SQL Oracle rounds a number down to the nearest whole number.

```
SELECT number, FLOOR(number) AS floor_value

FROM your_table;
```

The output of this query would look like the following:

NUMBER	FLOOR_VALUE
3.2	3
-3.2	-4
7.11	7
-0.01	-1

The **FLOOR** function can also be used with expressions, not just individual numbers. For example, **FLOOR**(column1 * column2) would first calculate the product of column1 and column2, then round it down to the nearest whole number.

If the number is already an integer, the **FLOOR** function will simply return the same number.

CEIL is **FLOORs** opposite, it stands for ceiling. It will round the number up to the nearest integer. You can use a **CASE** statement to define different rules for rounding based on whether the number is positive or negative.

For example:

```
SELECT CASE WHEN number >= 0 THEN FLOOR(number)
ELSE CEIL(number) END ;
```

This would round positive numbers down and negative numbers up.

GREATEST() and LEAST()

Return the greatest or smallest value of the list of arguments.

```
SELECT GREATEST(o.modified_dt, o.effective_dt)
FROM schema1.orders o;
```

The above query compares the modified_dt to the effective_dt and returns the latest of the two. If the modification date is null, it simply returns the effective date. The query below is similar except it will return the earliest of the two dates.

```
SELECT LEAST(o.modified_dt, o.effective_dt)
FROM schema1.orders o;
```

This query essentially retrieves the "earlier date" for each order. This allows you to identify the date that either represents the order's initial activation or its most recent update, whichever came first.

INITCAP()

Returns a value where the first letter in each word is uppercase, and all others are lowercase.

For example:

```
SELECT INITCAP(c.last_name)

FROM CUSTOMERS c;
```

This would return all last names with the first letter capitalized and the rest in lowercase regardless of how it appears in the data.

INSTR()

Searches for a value within another value. In other words, it searches a string for another substring and returns the starting position of the first occurrence of that substring. It's a great tool for extracting specific information from string data within your queries.

The syntax for **INSTR** is as follows:

```
INSTR(string, substring, start_position, occur-
rence)
```

String: The string to search in.

Substring: The substring to search for.

Start_position (optional): The character position within the string from where to start searching. Default is 1 (first character).

Occurrence (optional): The specific occurrence of the substring to find. Default is 1 (first occurrence).

INSTR is case-sensitive by default. If the substring is not found, **INSTR** returns 0. You can use -1 in the start_position to start searching from the end of the string backward.

For example:

To find the second occurrence of "a" in "Banana" starting from position 3:

```
SELECT INSTR('Banana', 'a', 3, 2)

FROM dual;
```

This would return a value of 6.

Or if you were looking for all the cases that contained the word “fraud” in a text field you could use the following:

```
SELECT INSTR(c.txt_field, 'fraud')

FROM schema5.comments c;
```

This would return 0 if the c.txt_field did not contain the word fraud and a number greater than 0 if it did.

LAST_DAY()

Returns the date of the last day of the month that contains date. The return type is always **DATE**, regardless of the datatype of date.

```
SELECT LAST_DAY(date)

SELECT LAST_DAY (SYSDATE)

SELECT LAST_DAY(created_date)
```

This can be a literal date value enclosed in quotes ('YYYY-MM-DD'), a column reference containing date values, or an expression resulting in a date value.

LENGTH()

In the **SELECT** statement **LENGTH** will calculate the length of a string stored in a column. In a **WHERE** clause **LENGTH** returns only rows with specified number of characters.

```
SELECT *

WHERE LENGTH (billing_cd) = 3

FROM table1;
```

The above query would return all billing_cd values that are only three characters in length.

LISTAGG()

The **LISTAGG** or list aggregator is an aggregate function that concatenates rows of text with a delimiter. It's commonly used to combine related values from a column into a single list for display or further analysis.

```
LISTAGG(expression, delimiter) WITHIN GROUP (ORDER BY
column)
```

Expression: The column or expression that contains the string values to concatenate.

Delimiter: The character that will be used to separate the string values. The default delimiter is a comma (,).

If you had a PRODUCT table with the following data:

PRODUCT_ID	PRODUCT_NAME	CATEGORY
1	Laptop	Electronics
2	Phone	Electronics
3	Shirt	Clothing
4	Dress	Clothing
5	Novel	Books
6	Textbook	Books

Then you would use LISTAGG to group laptop and phone into the same row because they have the same category. Shirt and dress would also appear on the same row as well, etc.

You would do this by using the following query:

```
SELECT p.category, LISTAGG(p.product_name, ',') WITHIN
GROUP (ORDER BY p.product_name) AS product_list

FROM product AS p

GROUP BY p.category;
```

The results of this query would look like this:

CATEGORY	PRODUCT_LIST
Electronics	Laptop, Phone
Clothing	Shirt, Dress
Books	Novel, Textbook

LPAD() and RPAD()

Adds the specified number and type of characters to the left of the specified value.

LPAD(column, length, pad_expr)

Column: The string you want to pad.

Length: The total desired length of the padded string.

Pad_Expr: (Optional) The string used for padding. If not specified, the default is a single space.

For example:

To pad a string of "SQL Tutorial" to 20 characters with dashes:

```
SELECT LPAD('SQL Tutorial', 20, '-') AS padded_string
FROM dual;
```

The output would look like the following:

--------SQL Tutorial

To pad a numerical value (123) to 10 characters with zeros:

```
SELECT LPAD(123, 10, '0') AS padded_number
FROM dual;
```

The output would look like the following: 0000000123

MEDIAN()

The **MEDIAN** function calculates the middle value of a set of numbers. This is a **GROUP BY** function. **MEDIAN** ignores null values within the group.

```
SELECT e.department, MEDIAN(e.salary) AS median_sal

FROM SCHEMA1.employees e

GROUP BY e.department;
```

This query calculates the median salary for each department in the employees table.

You can use the **OVER** clause with **MEDIAN** to calculate rolling medians within a partition (see page 109).

MONTHS_BETWEEN()

This function accurately calculates the number of months between two dates, including partial months.

For example:

You could use **MONTHS_BETWEEN** in the **WHERE** clause to find employees with tenures between 6 and 12 months:

```
SELECT employee_name, hire_date

FROM employees

WHERE MONTHS_BETWEEN(SYSDATE, hire_date) BETWEEN 6 AND
12;
```

You could also use **MONTHS_BETWEEN** to calculate the number of months between two given dates as in the following example.

```
SELECT MONTHS_BETWEEN
                    (TO_DATE('2024-06-30', 'YYYY-MM-DD'),
                     TO_DATE('2023-12-31', 'YYYY-MM-DD'))
AS months_difference

FROM dual;
```

This query returns 6.032258. This indicates there are approximately 6.03 months between the two specified dates.

You could use **MONTHS_BETWEEN** to calculate the average time between order placement and fulfillment:

```
SELECT AVG(MONTHS_BETWEEN(fulfillment_dt, order_date))
AS avg_time_to_fulfill

FROM orders;
```

You could use **MONTHS_BETWEEN** to get age:

```
SELECT TRUNC (MONTHS_BETWEEN (SYSDATE, c.birth_dt)/12)

FROM customer_info c;
```

In the above example the **MONTHS_BETWEEN** function calculates the number of months between two dates (the **SYSDATE** which represents the current system date and the birth_dt column.) Dividing it by 12 (the number of months in a year) will convert the number of months to years.

The **TRUNC** function removes the fractional part of the result, ensuring only whole years are considered. This method assumes birth date is stored in a date or timestamp data type.

Leap years are automatically accounted for when using **MONTHS_BETWEEN**.

NEXT_DAY ()

This function provides you with the next date that falls on a specified weekday after a given date.

The syntax for this function is:

```
NEXT_DAY(date, weekday)
```

For example:

```
SELECT NEXT_DAY(TRUNC(SYSDATE), 'SATURDAY ')

FROM dual;
```

This query would return the Saturday date that follows the date you ran the query.

NOT

Excludes anything meeting given condition usually used before a function. For example, **NOT IN, NOT EXISTS, NOT NULL, NOT BETWEEN, NOT LIKE**.

```
SELECT column1, column2, …

FROM table_name

WHERE NOT function condition;
```

NVL ()

This function is used to replace NULL value with another value.

```
NVL(column,0)
```

The above replaces all of your null values with 0.

NULLIF()

The **NULLIF** function compares two expressions and returns a null value if the expression is equal to the result. Otherwise, it returns the first expression unchanged. This essentially lets you eliminate rows where specific values match, making it a valuable tool for cleaning and filtering data.

For example, if you wanted the query to filter out customers on a specific date because of an unusual event or because it was an unusual day that you wanted to exclude you could use the following:

```
SELECT customer_id, last_purchase_dt

FROM customers

WHERE NULLIF(last_purchase_dt, '1JAN20') IS NOT NULL;
```

Because **NULLIF** is in the **WHERE** clause it ensures rows with null or '1JAN20' dates are excluded from the result set.

PIVOT()

The **PIVOT** operator in SQL is used to transform a table-valued expression by turning the unique values from one column in the expression into multiple columns in the output. The **PIVOT** operator also performs aggregations on any remaining column values that are wanted in the final output.

```
SELECT column_name, value1, value2, ...

FROM table_name

PIVOT (

  aggregate_function(column_name)

  FOR column_name IN (value1, value2, ...) );
```

If you had a table labeled **SALES** and it had the following data:

MONTH_YEAR	PRODUCT_ID	REGION	SALES_AMOUNT
1_2023	11	SOUTH	$1,500
2_2023	11	SOUTH	$2,500
1_2023	11	NORTH	$3,300
2_2023	11	NORTH	$2,700
1_2023	11	WEST	$2,300
2_2023	11	WEST	$7,700
1_2023	11	EAST	$4,350
2_2023	11	EAST	$5,200
1_2023	12	SOUTH	$2,300
2_2023	12	SOUTH	$7,300
1_2023	12	NORTH	$6,200
2_2023	12	NORTH	$4,000
1_2023	12	WEST	$1,500
2_2023	12	WEST	$2,500
1_2023	12	EAST	$3,300
2_2023	12	EAST	$2,700

Let's say you want to create a new table that shows the total sales for each product across all regions.

Here is the SQL query that uses the **PIVOT** function to achieve this:

```
SELECT s.product_id
     , 'NORTH' AS north_sales
     , 'SOUTH' AS south_sales
     , 'EAST'  AS east_sales
     , 'WEST'  AS west_sales
FROM sales AS s
PIVOT(
     SUM(s.sales_amount)
     FOR region IN ('NORTH', 'SOUTH', 'EAST', 'WEST')
     ) AS pivot_table;
```

The **PIVOT** function transforms the rows into columns creating the following data set.

PRODUCT_ID	NORTH_SALES	SOUTH_SALES	EAST_SALES	WEST_SALES
11	$6,000	$3,000	$9,550	$10,000
12	$10,200	$9,600	$6,000	$4,000

This table shows the total sales for each product across all regions, making it easier to compare sales performance for different products in different regions.

REGEXP_SUBSTR()

REGEXP_SUBSTR stands for "regular expression substring". It allows you to extract a specific substring from a string based on a defined regular expression pattern. It's a great tool for searching and manipulating textual data within your queries. Essentially it will return everything in a text string after or before a specified character.

The syntax for this function is as follows:

```
REGEXP_SUBSTR(column, character '[^start] N match_param
subexpr')
```

Column: This is the column from which the substring will be extracted.

Character: This is the regular expression that defines the pattern to match.

Start (optional): This specifies the starting position within the character to begin the search.

N (optional): This specifies the Nth occurrence of the pattern to extract. By default, the function extracts the first match.

Match_Param (optional): These are optional flags that can modify the behavior of the regular expression.

Subexpr (optional): This specifies which capturing group within the pattern to extract. By default, the entire matched substring is extracted.

For example, if you wanted to extract all the characters that follow the @ in a column containing emails you would use the following query.

```
SELECT c.email, REGEXP_SUBSTR(c.email, '[^@]+$') domain

FROM schema1.customers c;
```

In this example, the ^ is an anchor letting the query know you are about to enter the character to match against. The @ indicates the character in the string that the function will locate. Everything after the @ in the c.email column will be returned.

The + sign indicates that at least one character is required, but the search continues for further matches.

The $ sign indicates getting all the data from the character indicated to the end of the string. If the $ sign is removed then you will get all the data from the character indicated to the beginning of the string.

The data returned from the query above would look like this:

EMAIL	DOMAIN
MICKY@YAHOO.COM	YAHOO.COM
DONALD@GMAIL.COM	GMAIL.COM
GOOFY@AOL.COM	AOL.COM
MINI@YELP.COM	YELP.COM

To get everything before the @ in the c.email column delete the $ to get the following:

EMAIL	DOMAIN
MICKY@YAHOO.COM	MICKY
DONALD@GMAIL.COM	DONALD
GOOFY@AOL.COM	GOOFY
MINI@YELP.COM	MINI

To include the @ sign and stop at the "." Use the following:

```
SELECT c.email,REGEXP_SUBSTR(c.email, '@[^.]+') domain

FROM schema1.customers c;
```

The results would look like the following:

EMAIL	DOMAIN
MICKY@YAHOO.COM	@YAHOO
DONALD@GMAIL.COM	@GMAIL
GOOFY@AOL.COM	@AOL
MINI@YELP.COM	@YELP

To extract the first three words from a sentence:

```
SELECT REGEXP_SUBSTR(a.sentence, '\w+ \w+ \w+') AS
first_three_words

FROM schema1.articles a;
```

To extract the date from a sentence, where the dates format is dd-mm-yy (or 11-25-23):

```
SELECT REGEXP_SUBSTR(a.sentence, '\d2-\d2-\d2') AS
date

FROM schema1.articles a;
```

REPLACE()

REPLACE is useful for cleaning data, modifying text, and formatting output. It replaces all occurrences of a specified string or character within another string with a new string or character.

The syntax for replace is as follows:

```
REPLACE(char, search_string, replacement_string)
```

Char: The original string where you want to perform replacements.

Search_String: The string you want to find and replace.

Replacement_String: The string you want to substitute for each occurrence of the search string.

For example:

If you wanted to change the format of a phone number to remove the dashes, so that 987-654-3210 would change to 9876543210 you could use the following:

```
SELECT REPLACE (PHONE,'-','')
  FROM customers_info;
```

ROUND ()

The round function is meant for a numeric value.

```
SELECT ROUND(s.score, 2) FROM schema1.scores s;
```

This will round the score column to the second decimal place. So, if the data in the s.score column were 89.76593 the above query would return 89.77.

ROWNUM ()

ROWNUM is a pseudo-column (not an actual column in the table) that assigns a unique sequential number to each row in the result set. Its usefulness lies in using **ROWNUM** to limit the number of rows returned by a query.

For example:

```
SELECT *
FROM orders
WHERE ROWNUM <= 10;
```

This query will only show the first 10 rows from the orders table.

You can use **ROWNUM** in combination with an **ORDER BY** clause to order results based on their position within the result set.

For example:

```
SELECT *

FROM products

ORDER BY ROWNUM DESC ;
```

This query would list products in reverse order, or last row to first row.

It is important to note that **ROWNUM** values are not permanent identifiers for rows. They are assigned dynamically during query execution and can change if the underlying data or query order changes.

When using **ROWNUM** for filtering or ordering, ensure your query already filters down to the desired subset of data before applying **ROWNUM** limitations.

SUBSTR()

Returns a portion of the data in a cell beginning at the character position indicated at the indicated length.

```
SUBSTR(affected column, position in string,
        substring length)
```

An example of this would be the following:

```
SELECT SUBSTR(o.amt, 2, 3)

FROM schema1.orders o;
```

This query takes the o.amt column and starting at the 2nd character in the string it will grab the following three characters.

So, if the figure in the o.amt column was 987654321 the above query would capture 876.

SUM()

An aggregate function that is used to calculate the sum of all values in a column.

```
SUM(column)
```

The following query will calculate the total sales by summing up the values in the price column of the products table.

```
SELECT SUM(p.price) AS total_sales

FROM schema1.products p;
```

All aggregate functions require a **GROUP BY** statement if they are listed with other individual columns.

For example:

```
SELECT product_category, SUM(price) AS total_price

FROM schema1.products

GROUP BY product_category;
```

This query will calculate the total price for each product category. It will first group the products based on their product_category and then sum the values in the price column for each group.

SYSDATE

The **SYSDATE** function is a built-in function that returns the current date and time based on the server's operating system.

```
SELECT SYSDATE

FROM DUAL;
```

TO_CHAR()

Will convert a date, number, or expression to a character string with a specific format.

```
TO_CHAR(created_dt, 'Day') "Named Day of Week"

TO_CHAR(created_dt, 'D') "Numbered DAY OF Week"

TO_CHAR(created_dt, 'Month')"Named Month OF year"

TO_CHAR(created_dt, 'MM')"Numbered Month OF year"

TO_CHAR(created_dt, 'Q') "Numbered Quarter OF year"

TO_CHAR(created_dt, 'YYYY"Q"Q') "Year Quarter"

TO_CHAR(created_dt, 'Year')"Named year"

TO_CHAR(created_dt, 'YYYY')"Numbered year"

TO_CHAR(created_dt, 'dd-mm-yyyy') "Day-Month-Year"

TO_CHAR(created_dt, 'hh24:mi')"Time in 24hr format"

TO_CHAR(created_dt, 'iw-iyyy')"ISO Yr & Week of Yr"

TO_CHAR(created_dt, 'DL') "DATE IN LONG FORM"

TO_CHAR(created_dt, 'DS') "DATE IN SHORT FORM"
```

If you wanted to display a number with a currency symbol and two decimal places you would use the following syntax:

```
SELECT TO_CHAR(number_column, '$999.99') AS dollar_amt

FROM Table1;
```

While **TO_CHAR** is primarily used to convert numbers and dates to formatted strings, it can also be used to convert strings in specific situations.

```
SELECT TO_CHAR('2023-10-27', 'MM/DD/YYYY') AS format-
ted_date FROM Table1;
```

This query converts the string '2023-10-27' to a string formatted as '10/27/2023'.

You can also use **TO_CHAR** to extract characters from a string.

```
SELECT TO_CHAR('Product ID: 1234', '9999') AS prod-
uct_id FROM Table1;
```

This query extracts the digits from the string 'Product ID: 1234' and formats them as '1234'.

Another example of this is if you wanted to remove unwanted characters:

```
SELECT TO_CHAR('Phone number: (555) 555-5555', '999-
9999') AS phone_number

FROM Table1;
```

TO_DATE()

Converts a number or text string that represents a date into actual date and time values. The syntax for this is as follows:

```
TO_DATE(string, format)
```

If your data had the year and month displayed as 202010 then your equation would appear as follows:

```
SELECT TO_DATE(Year_mo, 'yyyymm')
FROM monthly_report;
```

TO_NUMBER()

Will convert a string to a numeric value.

The syntax for **TO_NUMBER** is as follows:

```
TO_NUMBER(string, 'format')
```

The following will convert a string column to a number.

```
TO_NUMBER (c.customer_id, '9999')
```

The following is an example of how to convert numbers with currency symbols.

```
SELECT TO_NUMBER('$125.00','L99G999D99')--Output: 125
     , TO_NUMBER('$123.00', '$9G990D00')--Output: 123
FROM dual;
```

The above query will convert the existing values into the given format, and make it an actual numeric value.

TRANSLATE()

The **TRANSLATE** function in SQL allows you to replace specific characters within a string with other characters.

The syntax for **TRANSLATE** is as follows:

```
TRANSLATE(string, characters, translations)
```

If you had a column that was an indicator that had both letters and numbers and you wanted to convert them into numbers to give them a ranking you could use the following:

```
SELECT TRANSLATE(w.schedule_ind,'FPSND24','1234576')

FROM SCHEMA1.workers w;
```

The schedule_ind is a column that contains a single character or numeric value. The possible values it could have include F, P, S, N, D, 2, 4. The above query lists those values, and then defines which values will replace them. So that any time the value equals F the **TRANSLATE** function will return a 1. If it is P then it will return a 2 and so on.

Another application of **TRANSLATE** would seek specific characters and change them to other characters. For example, if you had a table that stores phone numbers in various formats, including hyphens ("-") and parentheses ("()") around the area code. You want to standardize the format by removing all hyphens and parentheses, then you could use the following query:

```
SELECT customer_name

     , TRANSLATE(phone_number, '()-', '') AS stand-
ardized_phone

FROM customers;
```

Phone_Number is the string from which you want to remove characters. '()-' is the list of characters to be replaced. Single quotes are required around the list. '' is the empty string that will replace the listed characters. You can also specify other characters for replacement.

TRIM(), LTRIM(), and RTRIM()

TRIM Eliminates leading and trailing spaces.

```
SELECT TRIM (a.city) FROM addresses a;
```

This would turn a value ‘   York   ’   into ‘York’.

RTRIM removes trailing spaces.

```
SELECT RTRIM (a.city) FROM addresses a;
```

This would turn a value ‘York   ’   into ‘York’.

LTRIM is used to remove leading spaces from a string.

```
SELECT LTRIM (a.city) FROM addresses a;
```

This would turn a value ‘   York’   into ‘York’.

TRUNC()

TRUNC allows you to "cut off" unwanted parts of a number, date or time, resulting in a simpler representation.

The syntax for the **TRUNC** function is as follows:

```
TRUNC(date_time_expression, unit)
```

It is a best practice to always use trunc on date fields to get the full data on the given date range. The timestamp in the date field will often prevent you from getting all data in a given date range. **TRUNC** can eliminate the time stamp.

The query below demonstrates what is being excluded in a date range when trunc is not being used.

```
SELECT  P_KEY4, created_dt
FROM  SCHEMA_1.TABLE_1 c
WHERE TRUNC(c.created_dt) BETWEEN '2MAR2018' AND '3MAR18'
 MINUS
SELECT  P_KEY4,created_dt
FROM  SCHEMA_1.TABLE_1 c
WHERE c.created_dt BETWEEN '2MAR2018' AND '3MAR18';
```

The data set you would get in the above query command would look similar to the following:

P_KEY4	CREATED_DT
10003	3/3/2018
10004	3/3/2018

These are the records that are excluded because trunc is not used.

The data set you would get in the query before the **MINUS** command would look similar to the following:

P_KEY4	CREATED_DT
10001	3/2/2018
10002	3/2/2018
10003	3/3/2018
10004	3/3/2018

UPPER() or LOWER()

The **UPPER** function in SQL is used to convert all lowercase letters in a string to uppercase. Likewise, the **LOWER** function converts all uppercase letters in a string to lowercase.

```
SELECT UPPER(city) FROM addresses;

SELECT LOWER(EMAIL) FROM addresses;
```

Chapter 16: Complex Functions

The following functions, while powerful and versatile, necessitate a more comprehensive explanation than those covered in the Common Functions chapter. Their intricate details and nuances demand a deeper understanding to unlock their full potential.

These are some of the most useful functions for developers and analysts working with SQL in Oracle databases. By understanding and utilizing these functions effectively, you can significantly enhance your data manipulation capabilities, perform complex calculations, and extract deeper insights from your data.

1. The OVER (PARTITION BY) Function

The **OVER** clause is used to determine which rows from the query are applied to the function, what order they are evaluated in by that function, and when the function's calculations should restart. The **PARTITION BY**, much like a **GROUP BY** clause, is used to divide the result set into groups, or partitions, within a single column. This can be useful for performing calculations on individual rows with grouped data.

For example, (assuming you had multiple customers with multiple purchases) if you had a table that had a column for the customer name, item purchased, and sale price of item, you could use `OVER (PARTITION BY` to create a column that would give each row the total sum of purchases for that customer. So that every customer would have the total of all their purchases in every row with their name on it.

The syntax for this function is:

```
<function> OVER (PARTITION BY column
                           ORDER BY column)
```

```
SUM (column) OVER (PARTITION BY column
                          ORDER BY column)
```

To get a rolling (or cumulative) aggregate in a column leave out the **PARTITION BY** clause:

```
SELECT SUM(column2) OVER(ORDER BY column1)
```

The following is a chart of the various functions that can use the **OVER** clause:

R=Required; O=Optional; X=Not used

GROUP	FUNCTION	OVER clause	PARTI-TION BY	ORDER BY	ROWS or RANGE
Ranking	ROW_NUMBER()	R	O	R	X
Ranking	FIRST_VALUE()	R	O	R	O
Aggregate	SUM()	O	O	O	O
Aggregate	COUNT()	O	O	O	O
Aggregate	MAX()	O	O	O	O
Aggregate	MIN()	O	O	O	O
Aggregate	AVG()	O	O	O	O
Analytic	LAG()	R	O	R	X
Analytic	LEAD()	R	O	R	X

The following query has an example column for each type of window function listed above:

```
SELECT order_ID,customer_ID,gender_cd,date,Sale_amt
  , SUM(Sale_amt) over(ORDER BY date) Cumulitive_total
  , SUM (Sale_amt) over(PARTITION BY gender_cd ORDER
                         BY date) Cumu_Tot_by_GEND_CD
  , ROW_NUMBER() over(PARTITION BY gender_cd
                       ORDER BY date) count_GENDER_CD
  , FIRST_VALUE(order_date) over(ORDER BY customer_ID)
                                     First_order_date

  , COUNT(*) over(PARTITION BY customer_id ORDER
                         BY date) order_count
  , AVG (Sale_amt) over( ORDER BY date) Cumulitive_avg
  , MAX (work_week) over(PARTITION BY worker_id)
                                    AS Max_wk_worked
, LAG (Sale_amt) over(ORDER BY date) LAG
, LEAD (Sale_amt) over(ORDER BY date) LEAD
FROM SCHEMA_1.orders
WHERE Customer_ID = 78428
ORDER BY date;
```

This query provides a quick view, ten examples, of what the syntax might look like for each of the different window functions. We will now look at each of these functions individually and dive deeper into how each one works.

In the first example we used:

```
SELECT

SUM(Sale_amt) over(ORDER BY date) AS Cumulitive_total

, Sale_AMT, Date

FROM Orders;
```

This is adding a column to the query that provides a **SUM** of the SALE_AMT for every row providing a sum of that row and every previous row. So that each row will increase the total by the amount it provides. Output data could look like the following:

CUMULITIVE_TOTAL	SALE_AMT	DATE
100	100	1/1/2024
300	200	1/5/2024
350	50	1/6/2024
450	100	1/8/2024

In the second example we used:

```
SELECT

SUM (Sale_amt) over(PARTITION BY gender_cd ORDER
                     BY date)AS Cumu_Tot_by_GEND_CD
, gender_cd, Sale_AMT, Date

FROM Orders;
```

This is adding a column to the query that provides a **SUM** of the SALE_AMT for every row providing a sum of that row and every previous row by gender. So that each row will increase the total by the amount it provides according to the gender_cd. Output data could look like the following:

CUMU_TOT_BY_GEND_CD	GENDER_CD	SALE_AMT	DATE
100	M	100	1/1/2024
300	M	200	1/5/2024
50	F	50	1/6/2024
150	F	100	1/8/2024

The third example uses **ROW_NUMBER**.

ROW_NUMBER assigns a unique sequential number to each row within a result set. It is often used for tasks like pagination, ranking data, and identifying the first or last rows within a group. **ROW_NUMBER** operates on a set of rows (window) defined by a **PARTITION BY** clause (optional) and an **ORDER BY** clause (mandatory).

Thus, the example provided:

```
SELECT

ROW_NUMBER() over(PARTITION BY gender_cd
                  ORDER BY date) AS count_GENDER_CD
, gender_cd, date

FROM orders;
```

The **ROW_NUMBER** window function provides a column that provides separate counts based on gender_cd and in the order of the date. Output data could look like the following:

COUNT_GENDER_CD	GENDER_CD	DATE
1	M	1/1/2024
2	M	1/5/2024
1	F	1/6/2024
2	F	1/8/2024

The fourth example uses **FIRST_VALUE**.

FIRST_VALUE returns the first value within an ordered set of values. It's often used to identify initial values within groups or partitions, compare values across rows, or fill gaps in data. The **PARTITION BY** clause is optional for this function.

```
SELECT
FIRST_VALUE(order_date) over(ORDER BY customer_ID)
                                    AS First_order_date
, order_date

, customer_id

FROM orders;
```

The output data could look like the following:

FIRST_ORDER_DATE	ORDER_DATE	CUSTOMER_ID
1/5/2024	1/5/2024	1
1/5/2024	1/6/2024	1
1/5/2024		1
1/5/2024	1/9/2024	1
1/2/2024	1/2/2024	2
1/2/2024		2
1/2/2024	1/7/2024	2
1/2/2024	1/10/2024	2

The fifth example uses **COUNT**.

The window function **COUNT** will perform a count of the requested field for each item in the **PARTITION BY** statement.

```
SELECT

COUNT(*) over(PARTITION BY customer_id ORDER
                          BY date) order_count
 , date

 , customer_id

FROM orders;
```

Output data could look like the following:

ORDER_COUNT	ORDER_DATE	CUSTOMER_ID
5	1/1/2024	1
4	1/1/2024	2
3	2/1/2024	1
6	2/1/2024	2

The sixth example uses **AVG.**

The window function **AVG** will perform an average of the requested field separated by each item in the **PARTITION BY** statement if it is present, if it is not you will get a cumulative average as in the following example:

```
SELECT

AVG(sale_amt) over(ORDER BY date) Cumulative_avg

 , date

 , Sale_amt

FROM orders;
```

Output data could look like the following:

CUMULATIVE_AVG	ORDER_DATE	SALE_AMT
350	1/1/2024	350
354	1/2/2024	358
364	1/3/2024	384
356	1/4/2024	332

The seventh example uses **MAX.**

The window function **MAX/MIN** will return **MAX** or **MIN** of the requested field separated by each group in the **PARTITION BY** statement as in the following example:

```
SELECT

MAX (work_week) over(PARTITION BY worker_id)
                                    AS max_wk_worked
  , work_week

  , worker_id

FROM timecard;
```

Output data could look like the following:

MAX_WK_WORKED	WORK_WEEK	WORKER_ID
1/20/2024	1/6/2024	801385
1/20/2024	1/13/2024	801385
1/20/2024	1/20/2024	801385
2/3/2024	1/20/2024	801435
2/3/2024	2/6/2024	801435
2/3/2024	2/3/2024	801435

The eighth and ninth examples are **LEAD** and **LAG**.

The **LEAD** and **LAG** functions work within the context of a windowing clause, typically using **ORDER BY** to define the sequence of rows. They give you the value in the row after for **LEAD** and the value in the row before in the **LAG** function in the same result set.

By understanding and utilizing **LEAD** and **LAG**, you can delve deeper into sequential data within your Oracle databases, gleaning valuable insights and drawing connections between current and past values. They operate on ordered data, so ensure your data has a clear sorting order for accurate results.

This can be valuable if you are trying to compare one time period to another **LEAD** or **LAG** will let you but both periods in the same row so you can easily compare or perform a percent difference equation on them.

For example, say you wanted to compare the sales from the previous day to the sales of the current day. Your query might look something like this:

```
SELECT

LAG (Sale_amt) over(ORDER BY date) AS Prev_day_sale

 , Sale_amt

 , date AS order_date

FROM orders;
```

The output of this query would look like the following:

PREV_DAY_SALE	SALE_AMT	ORDER_DATE
	350	1/1/2024
350	358	1/2/2024
358	384	1/3/2024
384	332	1/4/2024

Notice the first row of the **LAG** column is NULL this is because there is no previous day in the data to pull an amount from. If it were the **LEAD** function then you would notice the last row would be null, because there is no future date to pull from to populate the last row with data.

2. The MAX() & MIN() Functions

These functions return the greatest or smallest value of the selected column.

When used without the **GROUP BY** clause, **MAX** and **MIN** will return a single row with the maximum or minimum value for the entire column. For example, if you have a table of employee salaries, and you run the query:

```
SELECT MAX(salary)

FROM schema.employees;
```

This query will return a single row with the highest salary in the entire table.

When used with the **GROUP BY** clause, **MAX** and **MIN** will return a row for each group in the result set. For example, if you have a table of employee salaries, and you run the query:

```
SELECT department, MAX(salary)

FROM schema.employees

GROUP BY department;
```

This query will return a row for each department in the table, with the highest salary for that department.

If you use **MAX** or **MIN** with multiple columns in the **GROUP BY** clause, you will get a row for each combination of values in those columns. For example, if you have a table of employee salaries, and you run the query:

```
SELECT department, job_title, MAX(salary)

FROM schema.employees

GROUP BY department, job_title;
```

This query will return a row for each combination of department and job title in the table, with multiple columns in the **GROUP BY** the highest salary for that combination.

A beginner mistake that is often made with the **MAX** and **MIN** functions, when they are used with multiple columns in the **GROUP BY** clause, is that they expect to see a row for each primary key but instead end up getting a row for each combination of values in those columns.

See the following example:

Wrong use of **MAX** if you only want one row per primary key (SSN in this example):

```
SELECT MAX(date)AS MAX, c.SSN , c.fil_cd, c.code_typ

FROM SCHEMA_1.TABLE_1 c

WHERE c.SSN = 123456789

GROUP BY c.SSN, c.CODE, c.code_typ;
```

This query will return a separate row for each unique combination of SSN, fil_cd, and code_typ. For example, if the individual with the SSN 123456789 has two different fil_cd values, and two different Code_typ values, then the query will return the max date for each unique combination of SSN, fil_cd value, and code_typ value.

Row #	MAX	SSN	FIL_CD	CODE_TYP
1	3/20/2023	123456789	W	1
2	2/12/2023	123456789	I	1
3	1/19/2023	123456789	I	2

Correct use of **MAX** if you only want one row per primary key (SSN):

```
SELECT c.date AS MAX, c.SSN , c.FIL_CD, code_typ

FROM SCHEMA_1.TABLE_1 c,

     (SELECT c.SSN , MAX(c.eff_dt) max

      FROM SCHEMA_1.TABLE_1 c

      WHERE c.P_KEY2  = 185364

      GROUP BY c.P_KEY2 ) s

WHERE s.P_KEY2  = c.P_KEY2

AND s.max = c.eff_dt

AND c.P_KEY2  = 185364;
```

This query will return only the row that corresponds to the maximum date:

Row #	MAX	SSN	FIL_CD	CODE_TYP
1	3/20/2023	123456789	W	1

Unless used as a double or single column, it is a best practice to use the **MAX/MIN** functions generally only as a sub query in the **FROM** clause as demonstrated above, or use the **OVER (PARTITION BY)** function to get your **MAX** or **MIN**.

3. CASE WHEN Statements

The **CASE WHEN** statement is SQL's way of handling if/then logic–SQL's equivalent of IF( ) function in Excel. The **CASE WHEN** statement is followed by at least one **THEN** statement. Every **CASE WHEN** statement must end with the **END** statement.
The **ELSE** statement is optional and provides a way to capture values not specified in the **THEN** statement.

```
SELECT p.player_name
       ,year
       ,CASE WHEN p.year = 'SR' THEN 'yes' ELSE 'no'
        END AS is_a_senior
  FROM benn.college_football_players p;
```

You can also define a number of outcomes in a **CASE** statement by including as many **WHEN/THEN** statements as you'd like:

```
SELECT c.player_name, c.weight,
CASE WHEN c.weight > 250 THEN 'over 250'
    WHEN c.weight BETWEEN 175 AND 249 THEN '175-249'
    ELSE '174 and under' END AS weight_group
  FROM benn.college_football_players c;
```

The results from this query would appear as follows:

ROW	PLAYER_NAME	WEIGHT	WEIGHT_GROUP
1	Bud Hodgson	257	over 250
2	Melissa Wiser	159	174 and under
3	Jack Frost	222	175-249
4	Patrick Black	270	over 250

In this example the **CASE WHEN** statement column will be labeled "weight_group" and the values that will appear in that column will be either "over 250", "175-249", or "174 and under."

The syntax for a **CASE WHEN** statement is generally as follows:

CASE WHEN (condition1) **THEN** (result1)
WHEN (condition2) **THEN** (result2)
. . .
ELSE (result_if_all_conditions_are_false) **END**

CASE: This keyword marks the beginning of the **CASE WHEN** statement.

WHEN condition **THEN** result: This clause specifies a condition to be evaluated. If the condition is true, the corresponding result is returned. If it is not true you get the result from the **ELSE** statement.

ELSE result_if_all_conditions_are_false: This clause provides an optional default result if none of the **WHEN** conditions are true.

END: This keyword marks the end of the **CASE WHEN** statement.

You can use functions in front of the CASE WHEN statement such as **SUM, COUNT, AVG.**

```
SUM (CASE WHEN c.gender_cd = 'F' THEN 1 ELSE null

END) total_female
```

Things to remember:

1. The **CASE WHEN** statement can handle multiple **WHEN** conditions, each evaluating a different condition.

2. The **ELSE** clause is optional, but it's recommended to provide a default result for cases where none of the **WHEN** conditions are met.

3. The **CASE WHEN** statement can be nested within another **CASE WHEN** statement to create more complex conditional logic.

4. The MOD () Function

The MOD function, also known as the modulo operator, is used to return the remainder of a division or NULL if the divisor is zero. It takes two arguments: the dividend and the divisor. The dividend is the number that is being divided, and the divisor is the number that the dividend is being divided by. In simpler terms, it tells you what is left over after you divide one number by another.

The syntax for the MOD function in SQL is as follows:

MOD(dividend, divisor)

- dividend: The number you want to divide.
- divisor: The number you want to divide by.

For example, the following query calculates the remainder of 11 divided by 3:

```
SELECT MOD(11, 3)

FROM dual;
```

This query will return the value 2, which is the remainder when 11 is divided by 3.

The MOD() function can be used in a variety of ways. For example, it can be used to:

a. Check if a number is even or odd. For example, the following query will return 1 if the number is odd and 0 if the number is even:

```
SELECT MOD(your_column, 2)

FROM TABLE1
```

Where your_column is a number column.

b. The MOD function can also be used to check for specific patterns in data. For instance, to select all records where the last digit of a value is 5, you would use the following query:

```
SELECT SELECT *

FROM your_table

WHERE MOD(your_column, 10) = 5;
```

This query selects all records where the remainder of dividing the value in the your_column column by 10 is equal to 5. Since the divisor is 10, this effectively checks if the last digit of the value is 5.

c. Generate random numbers. For example, the following query will generate a random number between 1 and 100:

```
SELECT MOD(DBMS_RANDOM.VALUE  * 100 + 1, 100)
FROM dual;
```

5. Random Sampling

The **DBMS_RANDOM.VALUE** function in Oracle is a function that returns a random number in a specified range. The syntax of the **DBMS_RANDOM.VALUE** function is as follows:

```
SELECT DBMS_RANDOM.VALUE(1, 100)
FROM Dual;
```

This will generate a number between 1 and 100 including decimal numbers. You can use the **DBMS_RANDOM.VALUE** function without specifying a range and only decimal values between 0 and 1 will return.

You can get a specific number of random rows from a dataset by using the following:

```
SELECT *
  FROM (SELECT *
        FROM SCHEMA_1.TABLE_1
        ORDER BY DBMS_RANDOM.VALUE)
 WHERE ROWNUM <= 60;
```

The following is an example of what the output of this query might look like:

Row	PKey	Name
7	753159	Bill Madison
38	891445	James Starks
47	688354	Katy Lincoln
103	744652	Abe Pace

6. Interval Random Sampling

One can get an interval sample using the **MOD()** function. The table that the mod is operating against must have a column with sequential numbers to reference. If the table does not have such a column you can create one with the following syntax:

```
SELECT *

FROM

(SELECT ROW_NUMBER()over (ORDER BY P_KEY4) ROW_NUM,
c.*

FROM SCHEMA_1.TABLE_1 c

) Sub_query

WHERE MOD(Sub_query.ROW_NUM,50) --Select interval(50)

=(SELECT ROUND(DBMS_RANDOM.VALUE(1,49),0)FROM Dual)
--Choses start point range (1-49)

FETCH NEXT 60 ROWS ONLY; --Choses sample size(60)
```

It is best to keep the starting point figure lower than the interval figure as the query won't know what to do if it is higher and will return nothing. Below is an example of what the output might look like.

Row	ROW_NUM	P_KEY4	NAME
1	41	159753	John Nielsen
2	91	178995	Wade Preston
3	141	165389	Kyle Knox
...	...	...	...
59	2991	186376	Francine Christiansen
60	3041	193764	Matt Neria

Chapter 17: Query Writing Best Practices

Writing a query involves constructing a set of instructions that communicate with a database to retrieve or manipulate data. The specific syntax and structure of queries vary depending on the database management system (DBMS) being used, but there are some general principles that apply to all DBMSs.

1. **Clarify the request and get all the specifics you can up-front.**
   You likely know the data better than the individual making the request. Make sure you ask specific and detailed questions so you can ensure you are getting the data how they need it.

2. **Identify the Tables and Columns you Need**
   Find the tables and columns you need to write the query. The ALL_TAB_COLUMNS table can be a useful resource for doing this.

```
SELECT a.owner, a.table_name, a.column_name

FROM all_tab_columns a

WHERE (a.table_name LIKE '%CUSTOMER%'

       OR a.column_name LIKE '%SALES_CD%')

AND owner IN ('SCHEMA_1', 'SCHEMA_2');
```

3. **Look at the data in each table you are going to use.**
    - Become familiar with the format of the data and how the data is being stored.
    - Become familiar with any indicators or codes and find out what they mean.
    - Identify the primary and foreign keys in each table.
        - When you join your tables what kind of relationship are you creating? (one-to-one, one-to-many)
            - It is always a good idea to check your joins and ensure you are getting the kind of relationship you are expecting.

4. **As you add new tables to your query ensure you join them.**

    Add one table at a time and run the query after each change.

5. **As you build your query run the query each time you add something new**

    This will ensure you are getting the data back that you are expecting.

        - Compare the row counts to see if they changed.
            - If they did ensure that you know why and are okay with the new count (or

are ok with excluding or including the rows that either disappeared or showed up)

- **Look for duplicates:** You can make the entire query a subquery and do a **COUNT** (*) on your unique key with a **HAVING COUNT** (*) > 1

For example:

```
SELECT customer_id, COUNT(*) AS count
FROM (query)
GROUP BY customer_id
HAVING count > 1;
```

This formula will group the query by customer ID and count the number of orders for each customer ID. It will then select all customer IDs where the count is greater than 1, indicating that there are duplicate records for that customer ID.

- Look for data that looks out of place (not what you would expect)
    - If aggregating look at the data behind the query and ensure it matches your expectations.

6. **Take a single unique key and look at the data with which it is associated.**

Compare it to the system. Compare it to a query looking at the unique key without any filters.

7. **Get your query reviewed**
   A coworker or another seasoned query writer can review your logic.

   or

   Write the same query in a different way to see if you get the same result.

Chapter 18: Querying Dates

Generating a Table with a Specified Range of Dates

You can generate a table with a specified range of dates using the all_tab_columns table which is a table in every database. This example creates a table of dates from January 1, 2023, to the current date or SYSDATE.

SYSDATE is a built-in function that returns the current date and time of the operating system on which the database server is running.

```
SELECT TO_DATE('01-JAN-2023', 'DD-MON-YYYY') +
       ROWNUM -1 Dates

FROM all_tab_columns

WHERE ROWNUM <= TO_DATE(TRUNC(SYSDATE))-TO_DATE(
      '01-JAN-2023', 'DD-MON-YYYY')+1
```

Grouping Aggregate Data by Time Period

1. Group by day

You can group data by day and count occurrences within each day.

```
SELECT Trunc(c.DATE) day_dt, COUNT (*) Total

FROM SCHEMA_1.TABLE_1 c

GROUP BY TRUNC(c.DATE)

ORDER BY TRUNC(c.DATE)
```

DAY_DT	TOTAL
1 JAN 2023	235
2 JAN 2023	349
3 JAN 2023	169
4 JAN 2023	283

...

2. Group by week

Grouping data by week (7 day period), starting from different days of the week depending on your start day.

- Use 'D' or 'DAY' to make the starting day of your 7 day period the first day of the week.

```
SELECT TRUNC(c.DATE, 'D') date, COUNT (*) total

FROM SCHEMA_1.TABLE_1 c

GROUP BY TRUNC(c.DATE, 'D')

ORDER BY TRUNC(c.DATE)
```

or

```
SELECT TRUNC(c.DATE, 'DAY') date, COUNT (*) total
FROM SCHEMA_1.TABLE_1 c
GROUP BY TRUNC(c.DATE, 'DAY')
ORDER BY TRUNC(c.DATE)
```

This query will give a weekly count of all the data in TABLE_1 with Sunday as the starting date as shown below.

DATE	TOTAL	
4 FEB 2024	3235	⟵ Sunday
11 FEB 2024	4349	⟵ Sunday
18 FEB 2024	1169	⟵ Sunday
25 FEB 2024	6283	⟵ Sunday
...		

- Use 'W' to make the starting day of your 7-day period the same day of the week as the first day of the month.

```
SELECT TRUNC(c.DATE, 'W') Week_DT , COUNT (*)total
FROM SCHEMA_1.TABLE_1 c
GROUP BY TRUNC(c.DATE, 'W')
ORDER BY TRUNC(c.DATE)
```

WEEK_DT	TOTAL	
1 FEB 2024	3275	⟵ Friday
8 FEB 2024	4319	⟵ Friday
15 FEB 2024	1199	⟵ Friday
22 FEB 2024	6273	⟵ Friday
...		

Since February 1st falls on a Friday each grouping will group the week from Friday to Thursday.

- Use 'WW' to make the starting day of your 7 day period the same day of the week as the first day of the year.

```
SELECT TRUNC(c.DATE, 'WW')Week_DT , COUNT (*) total

FROM SCHEMA_1.TABLE_1 c

GROUP BY TRUNC(c.DATE, 'WW')

ORDER BY TRUNC(c.DATE)
```

WEEK_DT	TOTAL	
5 FEB 2024	3239	⟵ Monday
12 FEB 2024	4342	⟵ Monday
19 FEB 2024	1167	⟵ Monday
26 FEB 2024	6284	⟵ Monday
...		

Since January 1st falls on a Monday each grouping will group the week from Monday to Sunday.

3. Group by Month

Grouping data by month.

```
SELECT TRUNC(c.DATE, 'MM') Month_dt, COUNT (*) total

FROM SCHEMA_1.TABLE_1 c

GROUP BY TRUNC(c.DATE, 'MM')

ORDER BY TRUNC(c.DATE, 'MM');
```

MONTH_DT	TOTAL
January 2024	10,235
February 2024	11,349
March 2024	9,169
April 2024	12,283
...	

4. Group by quarter

Grouping data by Quarter.

```
SELECT TRUNC(c.DATE, 'Q') Qtr_dt, COUNT (*) total

FROM SCHEMA_1.TABLE_1 c

GROUP BY TRUNC(c.DATE, 'Q')

ORDER BY TRUNC(c.DATE, 'Q');
```

QTR_DT	TOTAL
January 2024	30,753
April 2024	31,349
July 2024	29,169
October 2024	32,283

...

5. Group by Year

Grouping data by Calendar Year.

```
SELECT TRUNC(c.DATE, 'Y') Yr_dt, COUNT (*) total

FROM SCHEMA_1.TABLE_1 c

GROUP BY TRUNC(c.DATE, 'Y')

ORDER BY TRUNC(c.DATE, 'Y');
```

YR_DT	TOTAL
2021	130,753
2022	131,349
2023	129,169
2024	123,554

6. Group by Fiscal Year

Grouping data by fiscal year, starting from July 1st of the previous year to June 30th of the year listed.

```
SELECT CASE WHEN TO_CHAR(c.DATE, 'Q') IN ('3', '4')
      THEN TO_NUMBER(TO_CHAR(c.DATE, 'YYYY')) + 1
      ELSE TO_NUMBER(TO_CHAR(c.DATE, 'YYYY')) END FY
    , COUNT(*) total
FROM SCHEMA_1.TABLE_1 c
WHERE trunc(c.DATE) BETWEEN  '1jul21' AND '30jun24'
GROUP BY CASE WHEN TO_CHAR(c.DATE, 'Q') IN ('3', '4')
         THEN TO_NUMBER(TO_CHAR(c.DATE, 'YYYY')) + 1
         ELSE TO_NUMBER(TO_CHAR(c.DATE, 'YYYY')) END
ORDER BY 1;
```

FY	TOTAL
2021	120,753
2022	137,349
2023	131,169
2024	127,554

SQL Resources

https://www.w3schools.com/sql

https://mode.com/sql-tutorial/

https://sqlzoo.net/wiki/SQL_Tutorial

https://www.solo learn.com/Course/SQL/

https://www.udemy.com/introduction-to-oracle-sql/?ranMID=39197&ranEAID=JVFxdTr9V80&ranSiteID=JVFxdTr9V80-x0uDAeUBarSkolf_Y5mJPQ&LSNPUBID=JVFxdTr9V80

Practice

https://sqlzoo.net/

https://www.stratascratch.com/

https://datalemur.com/

https://leetcode.com/

Thank you for buying and using
SQL Basics: Datamining Reference Guide!

If you find this book useful, **please leave a review on Amazon.com.**

I would love to hear your feedback.
Thanks so much!!

~ Bradley L Salmond Jr

A special thank you to Patrick Frost who has been a good friend and partner in learning SQL for many years, and has no doubt contributed to many of the ideas set forth in this book.

Made in the USA
Monee, IL
03 November 2026

67380116R00085